GENETICALLY
ENGINEERED
ORGANISMS

GENETICALLY ENGINEERED ORGANISMS

BENEFITS AND RISKS

J. R. S. Fincham and J. R. Ravetz

in collaboration with a working party
of the Council for Science and Society

UNIVERSITY OF TORONTO PRESS
Toronto · Buffalo

Published in Great Britain by
Open University Press

First published in North America 1991
by University of Toronto Press
Toronto and Buffalo

ISBN 0-8020-5918-X (cloth)
ISBN 0-8020-6863-4 (paper)

Canadian Cataloguing in Publication Data

Fincham, J. R. S.
  Genetically engineered organisms

ISBN 0-8020-5918-X (bound).—ISBN 0-8020-6863-4 (pbk.).

1. Genetic engineering.   2. Biotechnology.
I. Ravetz, Jerome R.   II. Council for Science and
Society.   III. Title.

TP248.6.F5 1990   660'.65   C90-095602-X

Printed in Great Britain

CONTENTS

PREFACE

This book is the outcome of the activities of a Working Party convened by the Council for Science and Society with the initial purpose of looking at the problems associated with environmental release of genetically engineered organisms. The Group decided fairly early in its discussions to extend its remit so as to include other consequences of genetic engineering that are just as great a source of hopes and fears, particularly its present and potential application to medical diagnosis and therapy.

The introductory and two final chapters of the book were written mainly by Jerry Ravetz, and most of the remainder by John Fincham with first drafts of Chapters 3 and 4 provided by Simon Baumberg and John Beringer respectively. The whole Working Party reviewed the final draft and the book thus represents a consensus, in general if not necessarily in all details.

The Working Party were greatly helped by advice from Mr Brian Ager, Dr Julian Davies, Professor David Onions, Professor Bob Williamson and Professor Mark Williamson, all of whom gave generously of their time in travelling to meet the Group. These experts cannot, of course, be held responsible for any errors of fact that may have crept into the final text, nor for any of the opinions expressed.

The membership of the Working Party was as follows:

Dr Simon Baumberg, Department of Genetics, University of Leeds.

Professor John Beringer, Department of Microbiology, University of Bristol.

Dr Bernard Dixon, science writer and editor, Vice-Chairman of the Council for Science and Society.

Professor John Fincham, FRS (Chairman), Department of Genetics, University of Cambridge.

Professor Richard Flavell, Director, The John Innes Institute, Norwich.

Donna Haber, Regional Officer, Manufacturing, Science and Finance Union.

Dr Roger Land (deceased)†, Institute of Animal Physiology, Edinburgh Research Station.

Professor Kenneth Mellanby, Former Director, Monkswood Experimental Station.

Dr Jerry Ravetz, Chairman of the Council for Science and Society.

Mrs Felicity Sieghart, JP, formerly Chairman of the Council of the National Association of Gifted Children.

About the Council

The Council for Science and Society was established in 1973; it was among the pioneers in calling attention to the broader questions of scientific advance. Problems of the environment, health, technological risks and the consequences of scientific research are now of unprecedented public concern. To make its contribution to the discussion that a democratic society requires, the public must be properly informed. The media and the various interest groups provide for this, each in their own way. But there remains a need for information and analyses that are independent, authoritative and timely. The Council for Science and Society serves this purpose.

The Council has published a number of influential reports, including *The Acceptability of Risks, Superstar Technologies*,

---

† The Working Party was grieved to learn of the death of Roger Land on 17 April 1988.

Human Procreation, *New Technologies*, *UK Military R&D* and *Benefits and Risks of Knowledge-Based Systems*. It is now sponsoring studies on: Science in the Administration of Justice; Risks and Benefits of Genetic Engineering; Science and Sport; Food Safety; Sustainability, Science and Society; and the Human Genome Project.

The Council has a network of contacts in science, journalism, education, government and industry, who provide expert advice and assistance. Its work is done through public meetings, workshops and study groups. It publishes reports, occasional papers and press statements, and provides evidence to official inquiries. Support from several grant-awarding bodies, currently, including the Renaissance Trust, the Esmee Fairbairn Trust, the Wolfson Foundation and the Nuffield Foundation, enables the Council to maintain its independence.

The Council consists of some fifty men and women, who are prominent in science or some related area of public affairs. The Council's Chairman is Dr J. R. Ravetz, formerly Reader in the History and Philosophy of Science at the University of Leeds, and the author of *The Merger of Knowledge with Power*. The Vice-Chairman is Dr Bernard Dixon, former editor of *New Scientist* and author of *Science and Society* and *The Science of Science*.

Anyone interested in the Council's work is welcome to contact the office at 3–4 St Andrews Hill, London EC4U 5BY, Tel. 071-236-6723, Fax -7460.

1 | INTRODUCTION – HOW SAFE ARE ENGINEERED ORGANISMS?

The problem, and our approach

Every new technology brings its own mixture of promises and threats. With genetic engineering, the new powers to transform living organisms for the greater benefit of mankind are mirrored by the threatened violation of taboos about meddling with life itself. One feature of the present debate, that makes it so difficult to resolve, is that the operations that are now feasible are still relatively small scale and benign. Both proponents and opponents of the new technology are mainly arguing not in terms of what is, but of what might come some day soon. In this book, we have decided to keep to a narrow path, describing the state of the art as of now, on the possible threats and on the means to their containment. We hope that it will be a useful source of information and a commentary that will serve as a basis for further discussion and debate.

We have another function in mind for this book, which in our view is at least as important as that of an objective survey of the scene as at present. As the debate on genetic engineering develops (which it most certainly will), this report may serve as a sort of bench-mark, of where things were in 1990. As such, it will provide retrospective information on the pace of developments in the various fields.

The conclusions which we will draw are also framed with this dimension of time and development very much in mind. The task for society is not so much to achieve a perfect control at this moment over all possible experiments and developments, as to start to design a system of regulation (including directives, personnel and attitudes) which will be competent to steer the matured technology in beneficial directions.

There is no question of simply halting all developments in genetic engineering; if nothing else, it is now impossible to draw a workable distinction between these and 'conventional' methods. We do better to get some shape of the problem as it exists now, to monitor its changes over the coming years, and to act with conscious prudence to control any possible harm that may arise from these new developments.

Past experience – reassurances and cautions

Humans have for many years been releasing microbes into the environment, and have learned how to do so safely. Many pathogens can move from human wastes to drinking water and food. Modern sewage treatment has broken this cycle. Even now, sewage, discharged continuously and in huge quantities, contains enormous numbers of bacteria and other organisms – including many disease-causing varieties. Sewage treatment plants are by no means totally isolated from the environment, and considerable quantities of sewage sludge are dumped on to agricultural land in many countries. Yet normally this massive off-loading of microorganisms into our environment occurs without harm. Likewise, we have vast experience of both human and veterinary vaccines that have been genetically altered (though not, until very recently, by recombinant DNA techniques).

Moreover, *Homo sapiens* has long interfered with natural ecosystems and introduced 'foreign' species into new environments. The most relevant examples are of pest control. From the advent of crop cultivation 10 000 years ago, humans pulled weeds, picked insects off crops, and encouraged birds and insects that preyed on pests. More recently, microbes such as *Bacillus thuringiensis* have been cultivated and consciously released to combat other forms of life which are inimical to crops. Over 100 different pests have been effectively controlled by introducing and establishing a species which is a natural enemy of the nuisance species.

But there have been failures too, and adverse consequences when some organisms have been disseminated in new terrain either accidentally or purposely without adequate thought. The Australian subcontinent provides the most spectacular examples, as with the rabbit and the prickly-pear. In America, there was salt-cedar which sucked up the groundwater in arid lands, and more recently the Japanese vine kudzu which envelops and kills trees in parts of the Southern states. Invasions by pathogenic microorganisms to new populations have been even more catastrophic; thus syphilis was brought to Europe during the Renaissance, and a host of European diseases attacked native peoples in America and Australasia.

All such harmful events occurred in the absence of the knowledge and caution which are now taken for granted. Yet their example provides us with a reminder that there can be no absolute security in the case of biological innovation. Neither the biology of plants or microorganisms, nor the ecology of natural environments, can have such predictive power as to guarantee that any particular new introduction will be harmless.

Some examples and their lessons

Two specific case histories provide reassuring evidence against the likelihood of harmful consequences following the release of organisms created by gene splicing. The first concerns strains of *Rhizobium*, a bacterium which forms nodules on the roots of leguminous plants such as peas, clover and lucerne. In this symbiotic association, the rhizobia provide the plant with nitrogen which they have 'fixed' in an assimilable form from the atmosphere. Around the beginning of this century, microbiologists realized that they could encourage nodulation, and thus boost the supply of nitrogen to plants, by inoculating seeds with these naturally occurring organisms. Ever since then, an industry has existed to supply preparations to farmers.

Millions of hectares of land are now treated with rhizobia every year. In the USA alone in 1980, this involved some 4 million kg of inoculant, every gram of it containing over 100 million rhizobia. Despite this amount of material – and the presence of other, contaminating bacteria in the preparations used in many countries – no health or environmental problems seem to have followed the use of rhizobia. This indicates that the release of microbes *per se* into the soil need not be inherently dangerous or environmentally damaging.

Another significant lesson from our experience of rhizobia inoculation, particularly in the USA and Australia, is that the practice has been only modestly successful, because in many cases the percentage of nodules formed by introduced rhizobia has been small compared with those produced by indigenous strains. Although scientists have expended considerable effort in trying to select or breed more competitive rhizobia, and learned a good deal about the organism, they still know very little about how to improve its performance.

Genetic engineers may therefore find it difficult to produce particularly competitive strains of other microbes, which are less well understood. The ecological lesson is that an organism intro-duced into a new environment will persist only if it finds a suitable 'niche' – a set of favourable environmental conditions to which it is peculiarly well adapted.

The second historical story stretches back some 10 000 years, to the origin of maize. Quite different from its nearest wild relatives, teosinte and tripsacum, maize seems to have been our earliest and most important feat of plant breeding. Whereas teosinte and tripsacum and grasses produce seed in the tassel at the top of the plant, maize produces it seeds on bulky ears growing out of the stalk most of the way up.

Because it appeared in an evolutionary instant, maize is almost certainly a product of human intervention, based initially on the simple selection of seed for propagation and the intercrossing of odd-looking plants, before more sophisticated breeding tech-niques were applied. Corn cannot grow as a weed, and does not occur in the wild, because the kernels are so firmly attached to the ears that they do not fall off. It has no natural dispersal mechan-ism, therefore, and is entirely dependent on human beings for its survival. Even if an ear is left in a field and becomes buried in the soil, the plants are so crowded together when the kernels germinate that they generally do not flower.

Some scientists cite maize as a counter-example to the fear, voiced during the early years of the recombinant DNA debate, that artificially contrived organisms might reproduce uncontrollably. Others argue that the structural changes which have made it dependent upon humans are relatively unusual, and that no overall principles should be drawn from this experience. But there is general agreement that centuries of breeding, carried out with gradually increasing understanding, have not yielded plants capable of causing environmental damage. The fact that even today conventional plant breeding entails gene transfers far more indiscriminate and uncertain than the precise splicing now possible through recombinant DNA techniques, does suggest that such manipulations are unlikely to throw up unpredictably hazardous products.

Five problems of deliberate release

On the basis of both successes and adversities in the past, an ecological consensus is emerging about possible candidates for deliberate release which would require particularly careful scrutiny. Five problems can be identified. First, incursions such as those of ants and mongooses in Hawaii suggest that 'generalist' species, which are relatively unfastidious about both their food and their habitat, are more likely to cause problems than those with more specialized requirements. It could be argued that this category should include general soil-living bacteria into which genes coding for insecticidal toxins have been inserted.

Secondly, there may be harmful consequences when an organism is introduced to a new site where it is untroubled by its former enemies and thus totally free to prey on other creatures. One example came some years ago from Lake Gatun in the Panama Canal Zone, where a predatory fish, poecilopsis, escaped from a fish farm and exterminated half the population of some of the lake's fish species.

Thirdly, genetic changes may cause ecological changes – as with the *Spartina townsendii*, a species of cord grass, which originated from a natural cross in the Solent, England, about a century ago. Its ability to grow on barren tidal mudflats has made it both a pest, narrowing waterways and blocking harbours, and a treasured ally when planted to reclaim land and prevent erosion.

Fourthly, side-effects need to be anticipated. One of many case histories which underline this point comes from Hawaii, where the importation earlier this century of parasites to control various moths did more than precipitate the extermination of intended species. Destruction of native caterpillars also led to a dramatic decline in predators such as wasps of the genus *Odynerus* – and probably to the extinction of the country's insectivorous birds as well.

The fifth problem, and arguably one of the most difficult for regulatory bodies to contend with, is when an originally harmless organism, released in a slightly different environment, becomes a major pest. Examples include myxomatosis virus, which is responsible for only a minor disease in South American rabbits, but was lethal when first introduced into Australia (and then Europe), and *Legionella pneumophila*, a bacterium which occurs commonly in natural water systems but causes Legionnaires' disease when it has proliferated within and been disseminated as a fine mist through certain types of modern plumbing.

These last examples remind us how microorganisms can interact with their environments with much greater variety of response and effect, than higher plants and animals. They are more difficult to detect, and to recall when something has gone wrong, than ordinary plants and animals. Although no-one (except perhaps in the military context) will try to engineer a novel pathogen, we should never forget the capacity of living organisms, especially microorganisms, to surprise us when something changes in their relation to the environment.

Acknowledgement

This chapter is partly based on *Engineered Organisms in the Environment*, by Dr Bernard Dixon, published following the First International Conference on the Release of Genetically Engineered Microorganisms, Cardiff, Wales, April 1988.

2 | DNA – HOW IT WORKS AND HOW IT CAN BE MANIPULATED

Structure and replication

Within every living cell there is a vast amount of coded information that specifies the precise chemistry of the organism and its
pattern of development. This information is encoded in a type of
molecule called deoxyribonucleic acid (DNA). The structure of
DNA is that of a double helix – two long strands helically
interwound. Each strand consists of a sequence of four different
kinds of unit, each unit consisting of a base, which may be
adenine (A), guanine (G), thymine (T) or cytosine (C), joined to
a sugar (deoxyribose) which in turn is joined to phosphate. In the
DNA strand the deoxyribose of each unit is linked to the phosphate of the next unit along the line; the bases project laterally
from the chain of alternating sugars and phosphates (Fig. 2.1).

The critical feature of the double-stranded structure is that it is
held together by specific point-for-point pairing between the
bases: A pairs with T and G with C. Thus the sequence of bases
along one strand specifies the sequence along the other; for
example, G–C–A–A–T–C along one strand would imply C–G–
T–T–A–G along the partner strand. The mode of synthesis of
DNA in living cells ensures that the sequence is preserved from
one generation to the next. As the cell divides to make two
daughter cells, the two strands of each double helix are peeled

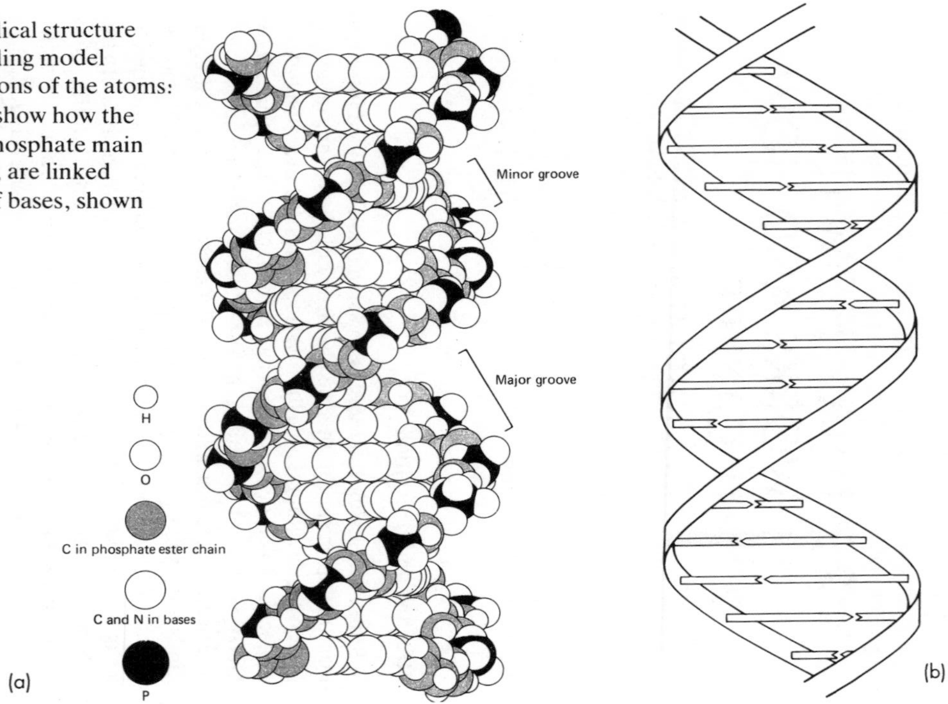

Fig. 2.1 The double-helical structure of DNA. (**a**) is a space-filling model showing the real dimensions of the atoms; (**b**) is a skeletal model to show how the helically wound sugar–phosphate main chains, shown as ribbons, are linked together by the pairing of bases, shown as rods.

Minor groove

Major groove

H

O

C in phosphate ester chain

C and N in bases

P

(a)

(b)

apart and a new partner is synthesized for each strand (Fig. 2.2). The sequence of each new strand is made to fit the sequence of the old strand so that the G–C and A–T pairings are restored.

Structural organization of DNA in cells

In bacteria and blue-green algae (prokaryotes) the whole of the basic genetic information (the genome) is contained in one to a few millimetres of double-stranded DNA in the form of a closed loop. To fit into the cell, the DNA loop is coiled and folded in association with basic proteins that serve to neutralize its acidity. It is usually referred to as the bacterial chromosome, a name borrowed from the functionally equivalent but differently organized DNA structures of eukaryotes (see below).

In addition to their standard genomes (chromosomes), bacteria commonly contain one or more kinds of much smaller closed-loop DNA molecules called plasmids. These provide the information for various functions, such as resistance to antibiotics, that can be regarded as 'optional extras' in that they are occasionally rather than always required for survival.

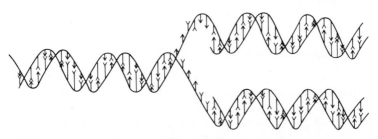

Fig. 2.2   Mechanism of DNA replication. The double helix unwinds and each strand serves as a template for synthesis of a new strand. DNA bases are represented as longer or shorter lines (purines and pyrimidines respectively) projecting inwards from the backbone deoxyribose–phosphate chains, and the complementary fit of adenine–thymine and guanine–cytosine base pairs is indicated by arrow heads and tails.

Higher plants and animals, as well as a wide range of simpler organisms other than bacteria, are included in the eukaryotes, the second great subdivision of the living world. Eukaryotes have the bulk of their DNA contained within a central cell compartment, the cell nucleus, and a minor fraction in other cell compartments (mitochondria and, in plants, chloroplasts). The DNA of a eukaryotic nucleus is always divided between a number of linear (not closed-loop) chromosomes, the number being characteristic of the species. In each chromosome a single molecule of DNA, which if fully extended would usually be of the order of centimetres in length, is packaged in a compact way with a characteristic array of proteins. Chromosomes of eukaryotes were originally defined as readily stainable rod-shaped structures, visible with the light microscope (chromosome means 'coloured body').

In terms of numbers of DNA base-pairs (bp), molecules of chromosomal DNA are of immense length. For this reason it is convenient to refer to DNA lengths in terms of thousands of base-pairs, or kilobases (kb). One millimetre of DNA corresponds to about 3000 kb. Each chromosome of a plant or animal contains a continuous DNA double helix with a sequence generally of the order of tens or hundreds of thousands of kilobases in length. However, only a fraction of this enormous length appears to consist of genes – the segments containing essential coded information.

Transcription of DNA into RNA

A gene is a unit which is transcribed into a form that can be translated into molecular structure in the rest of the cell. The transcript is another kind of nucleic acid molecule – ribonucleic acid or RNA – which is similar to a single DNA strand except that the sugar component is ribose instead of deoxyribose and thymine is replaced by a close analogue, uracil (U). The rules of transcription are similar to those of DNA replication in that the new RNA strand is complementary in base sequence to the DNA strand that served as the template. But, instead of remaining with the template as one strand of a new double helix, the RNA transcript is peeled off for use elsewhere in the cell (Fig. 2.3).

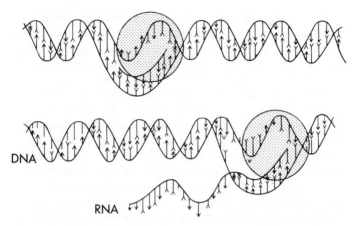

Fig. 2.3   Transcription of RNA from DNA. The enzyme RNA polymerase (represented by shaded circle) tracks along the double stranded RNA, inducing local unwinding of the double helix and transcribing one of the strands into RNA. Longer and shorter vertical lines represent purine and pyrimidine bases as in Fig. 2.2. The transcription start and stop points are signalled by special sequences in the DNA.

Transcription of a gene into RNA depends on a special sequence or set of sequences at the 'upstream' end of the gene to which the transcriptional machinery of the cell can bind and start copying. Special sequences of this kind are called promoters and these are often coupled to other sequences, called activators or enhancers, which increase their effectiveness. Without such sequences, appropriately positioned at one end of a gene, the information encoded within its sequence cannot be expressed.

Translation of RNA into protein structure

RNA molecules, transcribed from different genes, have various functions. The most numerous class, in terms of the number of different kinds if not in individual abundance, are messengers (mRNA), which are translated into the long chains of amino acids (polypeptides) that go to make up the very large number of specific proteins that are characteristic of each species of plant or animal. The translation proceeds by way of the general amino acid code, within which each sequence of three bases (with the exception of three triplets that act as stop signals) codes for one of the 20 different amino acids that occur in proteins. The linear sequence of bases in the mRNA determines the linear sequence of amino acid units in the protein, and this in turn determines the properties of the protein that is the ultimate gene product.

A recently discovered complication that perhaps should be mentioned is that most genes in higher plants and animals are much longer than is needed to code for the protein product, so that the RNA transcript has long tracts of irrelevant sequence (introns) that are spliced out before translation. Why genes have introns is one of nature's mysteries, though for some genes in some organisms control of splicing is one way in which gene activity is regulated.

The central component of the translational apparatus is the ribosome, which tracks along the mRNA strand synthesizing the amino acid chain as it goes (Fig. 2.4). The ribosomes consist of proteins, themselves the products of mRNA translation, and two very abundant kinds of RNA, called ribosomal RNA or rRNA, which are the transcripts of special genes that are present in multiple copies. Other kinds of RNA participate in the decoding process and yet others in the splicing-out of introns.

Reverse transcription

The original 'central dogma' of molecular biology was that the pathway of information transfer was exclusively from DNA to RNA to protein. More recent work has uncovered an important class of exceptions to this general rule. A large class of animal viruses, the retroviruses, have RNA, not DNA, as the genetic material in the infective virus particle. When the viral RNA gains entry to the host cell, it is reverse-transcribed into single-stranded DNA – that is to say, DNA is made on an RNA template rather than vice versa. The enzyme responsible, reverse transcriptase, is encoded in the virus genome and present in the infectious virus particle. The reverse transcript is made into double-stranded

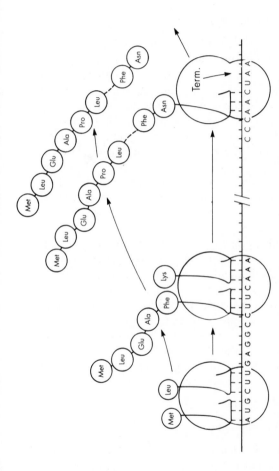

Fig. 2.4 Translation of messenger RNA (mRNA) into polypeptide chain of protein. Each sequence of three bases (codon) in the mRNA codes for an amino acid, except for three codons that signal chain termination. The ribosome reads off the mRNA base sequence in threes starting at a codon (AUG) for the amino acid methionine (Met), which is always the initiation codon. The codons are recognized by complementary sequences of three bases (anticodons) in molecules of transfer RNA (tRNA) which act as decoding keys. Each kind of tRNA attaches to one of the 20 different kinds of amino acid that make up polypeptide chains, and delivers that amino acid to the growing end of the polypeptide chain. Seven kinds of amino acid, represented by three-letter abbreviations, are represented in this diagram. When the ribosome reaches a chain-termination codon (UAA in the example shown), the completed polypeptide chain is released.

DNA by synthesis of a complementary DNA strand, and, by a breaking and resealing process inserted more or less at random into the host genome, where it can be transcribed back into RNA.

Control of the nature of the organism

DNA, through the processes of transcription and translation, is responsible for all the different kinds of proteins that are made in a particular organism. The proteins include both structural components of the cell and the specific catalysts (enzymes) of the chemical reactions of metabolism; together they determine the nature of the cell and how it interacts with other cells to build the whole organism. They also exert a number of feedback effects on the DNA that specified them. The replication of the DNA is catalysed by specific enzymes and protein co-factors. The timing and level of gene transcription is controlled by proteins which bind to gene promoters and associated activator/enhancer DNA regions. Thus there are genes that specify proteins that control the activity of other genes.

The whole complex programme of organism development is largely a matter of the right set of genes being expressed in the right place at the right time and at the right level. The programme is played out through a network of interactions of immense complexity which is being slowly unravelled but may never be fully described for any organism. We can be sure, however, that the programme is written in the DNA. Changes (mutations) in the coding sequences in the DNA will be heritable and have effects on the organism that may range from trivial to profound depending on the nature of the change and the function of the gene in which it occurs.

The advent of DNA manipulation

The gene concept was founded on the basis of studies of the patterns of transmission through successive generations of inherited differences between individuals of a species. The differences concerned were either found in nature or were the results of mutations induced by various treatments, including X-irradiation and exposure to mutagenic chemicals. But whatever the origin of the variation it was essentially random so far as the geneticist was concerned; he had to take what nature or random mutagenesis gave him, and if he wanted a special kind of mutation he had to hunt for it among large numbers of mutagenized individuals. Although there were good reasons for believing that the genes were made of DNA, there was no way of isolating any particular gene as a separate chemical entity.

Molecular probing for genes

In the 1970s new techniques were developed that allowed the physical detection of specific genes. First, biochemists began to isolate specific mRNA molecules which, because of their complementary sequence relationship to the genes from which they had been transcribed, could be used as molecular probes. Labelled with radioactive phosphate, an mRNA strand, or a complementary DNA (cDNA) strand reverse-transcribed from it with retroviral reverse transcriptase, can be hybridized to one or other of the two strands of the corresponding gene which thus becomes labelled with radioactivity (Fig. 2.5).

Reconstructing DNA molecules

The second breakthrough was the discovery of a class of bacterial enzymes, called restriction endonucleases, that cleave double-stranded DNA at specific sequences (a different sequence for each enzyme) and so cut out genes, or pieces of genes, within DNA fragments (restriction fragments) of characteristic sizes. An important property of restriction fragments is that, when derived by cutting with the same enzyme, their ends are mutually cohesive.

The third essential step was the discovery of an enzyme, DNA ligase, that would join mutually cohering restriction fragments end-to-end with stable chemical bonds. Thus one can piece together new DNA molecules, uniting fragments derived from different sources.

Cloning of DNA

The term 'cloning' means propagation without genetic change. It applies, for example, to propagation of plants by cuttings or

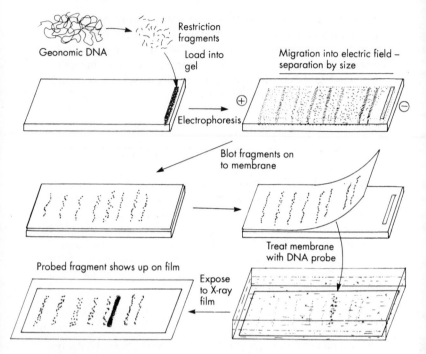

Fig. 2.5 Identifying a specific DNA sequence in the genome. Genomic DNA is cut with a restriction enzyme, and the fragments so generated are separated by size by electrophoresis in a gel. Being negatively charged, the fragments migrate toward the positive electrode; smaller fragments migrate faster than larger ones because of the sifting property of the gel. The DNA pattern following electrophoresis (with only highly repetitive sequences forming bands that stand out above the background smear) is transferred by blotting to a membrane to which the DNA becomes fixed. After treatment with alkali to separate the double-stranded DNA into single strands, the membrane is immersed in a solution containing radioactively labelled single strands of a fragment of cloned DNA (probe), which bind by complementary base pairing to homologous sequences fixed to the membrane. The radioactively labelled bands can then be located by exposure of the membrane to X-ray film.

tubers. By extension, it has come to be applied to the propagation of particular DNA molecules, making use of their capacity for accurate self-replication. The possibility of cloning DNA sequences arose from the study of bacterial viruses (bacteriophages) and plasmids. Both plasmid and phage DNA is endowed with the ability to replicate independently of the main bacterial chromosome. Phage or plasmid DNA can be readily purified and cut with a suitable restriction enzyme, and then any foreign DNA fragment cut out by the same enzyme can be joined into it. If the virus or plasmid DNA is then reintroduced into the bacterium, and means for doing this have been well known since the 1960s, the foreign DNA will be replicated in the bacterial cell along with the plasmid or virus DNA to which it is joined. In this way any DNA fragment can be selectively amplified within the bacterial culture, often to a very high concentration, and subsequently easily purified. Plasmids or phages used in this way are called cloning vectors, and they can be tailor-made for the purpose. It is often convenient to use a plasmid containing one or more genes, generally inserted artificially, conferring resistance to antibiotics. This enables easy selection of bacteria harbouring the plasmid.

The principle of cloning DNA fragments in a plasmid vector is illustrated in Fig. 2.6.

Gene banks

As a result of new technology it is now possible to isolate, as a specific sequence of DNA, any gene which can be identified through its effect on the host organism, or through the availability of a specific molecular probe. The general strategy is to make a gene 'library' or 'bank' by cloning a near-comprehensive set of large DNA fragments from the species under investigation into a suitable plasmid or phage vector, and then to screen the library with a probe or through some functional test to identify the gene that is being sought. Very many genes have now been cloned, especially from simple organisms such as yeast but to a large extent from higher plants and animals also. The most commonly used vectors are plasmids, almost always reconstructed to maximize their suitability to the purpose. The universally used

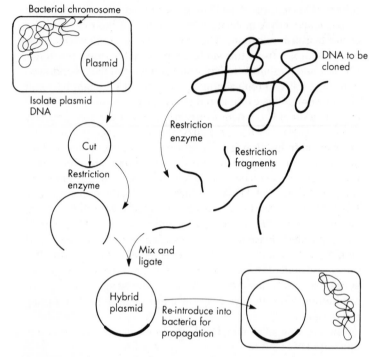

Fig. 2.6 One way of cloning DNA fragments. A fragment cut out of the genomic DNA (shown in thick line) with a specific restriction enzyme, and ligated into the closed-loop DNA (thin line) of a bacterial plasmid cut with the same restriction enzyme. The hybrid DNA construction is reintroduced into bacterial cells, where the inserted fragment is multiplied indefinitely along with the plasmid.

bacterial host is a harmless strain of the common gut bacterium *Escherichia coli*.

Early concern about recombinant DNA and the response to it

The early experiments on the cloning of DNA segments into artificial vectors were undertaken with little or no concern about possible hazards. Such exercises as the introduction of a frog gene into phage DNA for propagation in *E. coli* suggested no obvious scenario of risk. The first projected experiment to arouse concern in some of those participating in it was one in which it was proposed to introduce the DNA of a monkey virus (SV40) into *E. coli*. The worry about the possibility that this might be creating an opportunity for the virus to infect humans led to the postponement of this experiment and to further thought about possible hazards associated with recombinant DNA in general. The idea dawned that any organism carrying artificially reconstructed DNA could be a novelty of a kind that would be unlikely to arise naturally, and that it could therefore have unprecedented and perhaps dangerous properties.

Thus concern about the possible consequences of playing recklessly with known pathogens was joined to a general fear of the unknown that was not entirely irrational, since the unknown in question had the potential for unlimited replication.

These concerns led in 1974 to a call from a committee of the US National Academy of Science – a committee composed mainly of the scientists most actively involved – for a moratorium on recombinant DNA experiments until the whole issue had been debated. The debate, at that stage confined to the molecular biology community, took place at Asilomar, California, in 1975. A statement eventually agreed at that meeting called for the establishment of guidelines to regulate recombinant DNA research, and for the development of safe bacteria and safe vectors that would be incapable of surviving outside very special environments provided for them in the laboratory. In the same year the Ashby Committee set up in the UK made recommendations very much in the same spirit – that research involving recombinant DNA should be pursued but be subject to special regulation.

The scientists' proposals for regulation were, in 1976, taken up by governments and government agencies. In the USA, the Recombinant DNA Advisory Committee (RAC) of the National Institutes of Health drafted safety guidelines that had virtually the force of legislation. In Britain the same job was done by the Williams Committee [1], whose scheme of regulation came to be administered by the Genetic Manipulation Advisory Group

(GMAG), which derived its authority from the Health and Safety at Work legislation.

The rules adopted were essentially similar on both sides of the Atlantic. Requirements were laid down for different levels of containment of proposed experiments, ranging from grade 1, which was essentially standard microbiological sterile technique with restricted access to the laboratory, to grade 4, which involved all the protective apparatus – protective clothing, closed cabinets accessible only through glove boxes, air locks, and so on – that had hitherto been thought necessary only for the most dangerous pathogens. Experiments thought to require the higher levels of security were those that involved DNA of known pathogens or the random cloning of DNA from animals; the closer the animal appeared to be to man the higher the degree of risk attributed to the DNA, because of the chance that it might harbour latent virus DNA capable of causing human disease. Experiments with cloned DNA of apparently harmless organisms introduced into other apparently harmless species, or even into the same species, were placed into one of the lower categories, but were still subject to regulation at some level.

While all of this was going on the question of the safety or otherwise of cloned DNA entered the public domain. Far from being reassured by the fact that the scientists most involved in the work were acting in advance to head off possible danger, many concerned laymen reacted with alarm. They reasoned that if even the molecular biologists were worried, with their obvious vested interest in the continuation of the work, the dangers must be much greater than had been admitted. There was a period, roughly from 1977 to 1979, when DNA cloners were very much on the defensive [2], but they were still able to carry on with their work. The statutory regulations governing experimentation with recombinant DNA' were, on the whole, carefully adhered to; though somewhat troublesome and expensive, they were not found too inhibiting.

The present situation and new concerns

Over the past decades, the cloning of genes and their transfer between organisms has become commonplace in very many laboratories throughout the scientifically developed world. The object, until recently, has mainly been to analyse the working of existing kinds of organisms rather than to create new ones. Gene cloning, followed by DNA sequence determination and decoding, is the quickest way of determining the amino acid sequences of the proteins that the organism can make.

Structure does not in itself necessarily lead to understanding of function. But in experimentally favoured organisms (*E. coli*, yeast, fruit flies, mice) cloned DNA sequences can be reintroduced into living cells having been subjected to controlled modifications, and the effects of the changes then observed. In this way it is possible to determine the functions that proteins perform and the parts of their structures that are essential for those functions. It is because of its immense analytical power that DNA manipulation has become such a central activity over such a broad span of biology. If one scans the contents of journals of biochemistry or molecular biology or cell biology, one finds that almost a majority of papers involve genetic engineering, as it has come to be called.

Amid all this successful use of recombinant DNA technology, anxiety about possible risks arising from its use in the laboratory has receded. Realistic scenarios of risk attaching to recombinant DNA as such have not been forthcoming, and it has become rather generally accepted that what matters are the demonstrable properties of particular DNA sequences rather than the procedures used in their derivation. There has been a decade of intense activity in gene cloning without any clear example of a harmful accident. While it might be argued that this good safety record has been due to meticulous adherence to safety regulations on the part of everyone concerned, the absence of alarms has had a very reassuring effect. In this atmosphere, safety regulations governing most kinds of work with DNA have been substantially relaxed. The emphasis is now on self-assessment and *post-facto* notification except for experiments – for example those involving manipulation of dangerous pathogens – placed in the highest categories of risk.

Today, however, there is a new reason for concern. Organisms carrying artificially recombined DNA sequences are now finding uses outside the confines of the research laboratory – in pharmaceutical industry, in agriculture and in medicine. This is provoking fears about possible escapes of engineered organisms or viruses into the general environment and the possible consequences of such escapes for ecology and/or health.

In addition to ecological concern, there is some fear that, as DNA technology gets into the hands of the medical profession, the human genetic endowment – that most sacred of personal possessions – may be tampered with in ways that may be injurious to people's sense of identity. These fears are most powerfully expressed by the Green parties of continental Europe, where in some countries they are already strong enough to influence national policy, and they are shared by some radical groups in the USA. In the UK, the environmental issues have been addressed by the Royal Commission on Environmental Pollution who have very recently published a well-documented and balanced Report [3]. Regulations, proposed in this Report, for governing release of engineered organisms are included in the Environmental Protection Bill, due to receive the Royal Assent in November 1990.

It is our purpose in the following chapters to present an independent view, considering in some detail what is being and can be done through recombinant DNA techniques in industry, agriculture and medicine, and the importance of these actual and potential applications for our society.

References

1 *Williams Working Party Report* (1976) HMSO Cmnd 6600.
2 Watson, J. D. and Tooze, J. (1981) *The DNA Story*, 604 pp. W. H. Freeman, San Francisco.
3 Royal Commission on Environmental Pollution (Lord Lewis, Chairman) (1989) Thirteenth Report: *The Release of Genetically Engineered Organisms to the Environment*, x + 144 pp. HMSO.

3 GENETICALLY MANIPULATED ORGANISMS IN INDUSTRY

From being primarily research tools, genetically manipulated microorganisms are now beginning to be used in industry, mainly for the manufacture of pharmaceuticals for human and veterinary use [1]. This holds out the hope of cheaper and perhaps purer products. At the same time, the culture of genetically engineered organisms on an industrial scale may arouse fears of increased chances of escape into the environment and of possible risks for industrial personnel. In this chapter we describe the kinds of organisms and processes that are already being used or seem likely to come into use, and then discuss the hypothetical risks and the safeguards required to deal with them.

What products can be made by genetically manipulated organisms?

DNA for diagnostic purposes

Cloned DNA sequences from pathogenic bacteria or viruses can be used as diagnostic reagents. They can be used as hybridizing probes for the corresponding sequences of the pathogenic organisms, supposing that these have been isolated from the patient in sufficient quantity. A far more sensitive diagnostic method is to use short cloned sequences in pairs to prime the polymerase chain reaction, which is capable of amplifying a tiny amount of a particular DNA sequence to an easily detectable level (see Fig. 6.3, p. 78). There is a large market for diagnostic kits based on these principles.

Hormones and growth factors

Most of these substances are short-chain polypeptides or proteins, the latter being formed by folding and packing together of longer polypeptide chains. For the sake of simplicity we shall refer to all of these products as proteins, regardless of their chain lengths. They are direct gene products, and the genes encoding them can be cloned and introduced, for high-level transcription and translation, into bacteria or (for some purposes) animal or yeast cells.

Products already being made include human insulin (marketed in 1984), α_2-interferon (1985), human growth hormone (1986), tissue plasminogen activator, TPA (1988) and erythropoietin (1988–89). Insulin is required by diabetics and growth hormone by children congenitally deficient in it. α_2-Interferon is used in the treatment of certain kinds of leukaemia and serious liver infections by viruses hepatitis B and C. TPA is expected to be increasingly used to treat some sorts of heart disease by stimulating the dissolution of blood clots. Erythropoietin stimulates red blood-cell maturation and is intended for use in the treatment of anaemias and in connection with kidney dialysis. Among products under development, lymphokines (factors involved in normal lymph-cell maturation) are used, or could be used, to promote wound healing and as adjuvants to improve the efficacy of immunization; they are thought to have a particularly large potential market.

Each of these proteins can be produced in bacteria, yeast or cultured animal cells by adding its gene to the thousands already present. The desired gene is obtained, often as complementary DNA (cDNA, the DNA equivalent of the messenger RNA, see p. 11), and spliced into a DNA vector that will enable it to replicate in the cell of choice. The splicing is arranged so as to place the newly introduced gene immediately 'downstream' of

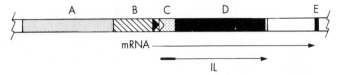

Fig. 3.1   An example of a patchwork DNA construction designed to confer upon yeast cells the ability to make the human hormone interleukin-ß and secrete it into the growth medium. In this example, the construct included, in upstream-to-downstream order, (A) an activator sequence from yeast with the effect of enhancing transcription on a certain kind of growth medium, (B) a strong yeast promoter region including a transcription start signal (black arrowhead), (C) a DNA sequence from a different yeast species encoding a short secretion-signal protein segment with its own translation initiation codon (open arrowhead), fused to (D) a human DNA sequence encoding interleukin-β and with a translation stop signal at the downstream end, and finally (E) a transcription-termination signal from the yeast plasmid in which the whole construction is cloned. When introduced into yeast cells the hybrid plasmid replicates and the patchwork construct is transcribed into messenger RNA (thin arrow) which is translated into interleukin-ß polypeptide (IL) with the signal sequence attached at its upstream end, ensuring that it is secreted into the growth medium. After Baldari *et al.* [2].

promoter (and sometimes enhancer) signals that will ensure its hyper-efficient transcription and subsequent translation into polypeptide chains. It is also possible to include in the construction a segment of DNA encoding a signal sequence in the protein product that will ensure its efficient secretion from the cell into the culture fluid. Protein is generally much easier to purify if it is already separated from the cell that produced it. An example of a patchwork DNA construction designed for protein secretion as well as for high productivity is shown in Fig. 3.1.

For ease both of genetic manipulation and cultivation of the cells, the organism most often used has been the harmless K-12 strain of the bacterium *Escherichia coli*. However, when the activity of the product is dependent on the addition of sugar (glycosyl) side-chains to the polypeptide chain, it may be essential to use eukaryotic cells, since only they possess the biochemical apparatus necessary for these secondary modifications of the gene product. Yeast is the most convenient eukaryote for genetic engineering, but, since yeast and animal patterns of glycosylation are not necessarily the same, a yeast product is not always an adequate substitute for one made in animal cells (or in intact animals – see Chapter 6, p. 80).

As regards the comparative cost of human proteins produced by genetic engineering as against their counterparts from conventional sources, the only comparison available so far is for insulin. At present the engineered product probably has no price advantage over the product isolated from pigs. However, this situation could change as the market becomes more competitive.

Antibodies and antibody subunits

The antibody molecules that are produced by mammals in response to challenge by foreign proteins (antigens) each normally consist of four polypeptide chains, two larger ('heavy') and two smaller ('light'). Each chain has a constant domain shared by a whole class of antibody molecules and a variable domain that is specific to the particular antibody. The variable region is the product of a type of DNA rearrangement, partly controlled and partly random, that occurs in the precursors of antibody-producing cells during their maturation, and it makes a precise fit to the target protein. Binding of antibodies, through their variable domains, to pathogen antigens is the first step towards the elimination of a pathogen from the system.

Antibodies directed against particular pathogens can now be fabricated in microbial cells engineered to produce them. A problem with the therapeutic use of engineered antibodies is that the genes encoding them are usually cloned from immunized mice; a mouse antibody is recognized as a foreign protein by the human immune system and would thus be likely to produce an immunological shock reaction in a human patient. The solution to the problem may be to 'humanize' the antibodies by using hybrid genes for their production. The part of the gene encoding a constant domain (about three-quarters of the total for heavy chains) can be cloned from human cells and spliced *in vitro* to a specific variable sequence from mouse [3].

It has recently been found that artificially produced variable domains of heavy chains will, by themselves, bind with high affinity to the corresponding antigens. This discovery [4] offers the possibility of engineering the production of single kinds of polypeptide chains that can be used either as diagnostic reagents for identification of particular pathogens or, more ambitiously, for therapy.

The development of tailor-made antibody polypeptides is still in its infancy, but it is clearly possible in principle not only to join human constant to mouse variable domains but also to improve affinity for the target molecules by detailed modifications of crucial codons by biochemical manipulation of the DNA clones. This is an area that may have enormous potential.

Antibiotics

These are molecules of relatively small molecular size, which are products of protein (enzyme) activity rather than proteins themselves. Many such compounds are already made industrially in large-scale cultures of microorganisms that produce them in large amounts, either naturally or because they have been subject to mutation and selection.

In some cases it has already been shown by experiments on a laboratory scale that further improvements can be made by genetic manipulation. The problem is inherently more complicated when one is dealing with a substance, like an antibiotic, that is the product of a chain of reactions each requiring a different enzyme protein, than when the desired product is just a single protein. But if one step in the synthesis is rate-limiting, it may be possible to get useful increases in yield by boosting the level of the enzyme responsible for that step. And it may even be possible to introduce multiple copies of the whole set of genes encoding all the enzymes of the pathway. Since some such genes are clustered in their organism of origin (e.g. the penicillin-synthesizing genes in fungi) it may be possible to clone them all together.

A further possibility that is currently being widely pursued is the use of genetic engineering to make new 'hybrid' antibiotics [5]. Related antibiotic-producing microorganisms often provide variations on a basic molecular theme. Thus organism A may make a molecule with a modification in one position and organism B a modification in another position. By recombining genes from A and B it may be possible to make a new organism that makes both modifications to yield a new antibiotic that could have useful properties.

Food additives and dietary supplements

We deal here with such small molecular substances as essential amino acids and vitamins which have to be supplied in the human diet. Because not all diets contain these substances in adequate amounts, there is a market for them, especially the vitamins, as additives and supplements. These are already being produced on an industrial scale by microorganisms, often members of the bacterial genera *Corynebacterium* or *Brevibacterium*. Much the same considerations apply as in the case of antibiotics, in that any improvement in yield would depend upon improving the overall efficiency of a multi-step metabolic pathway. It may be profitable to switch production from the bacteria hitherto used to species more developed for sophisticated genetic engineering, such as *Escherichia coli* and the yeast *Saccharomyces cerevisiae*.

Safety considerations

Are the products of genetic engineering safe to use?

Some people appear to fear all products of genetically engineered organisms as 'unnatural'. The recent furore in Europe over the import of American meat containing traces of injected bovine growth hormone (BGH, bovine somatotropin) seems to have been due at least in part to the belief that BGH, injected into cattle to increase meat or milk production, had been genetically engineered. In fact, the hormone was chemically identical to the hormone that cattle make for themselves; the artificial boost only increased the concentration of something that was already there. That the BGH may have been made by a genetically engineered microorganism should have been irrelevant – a molecule of a particular chemical structure is the same whatever its origin.

That said, it is fair to point out that not all pharmaceutically active proteins made by genetic engineering are quite the same as the corresponding natural products. For example, as noted above, many proteins carry precisely positioned sugar (glycosyl) side-chains, and these may modify their properties. Glycosylation of a mammalian protein will occur normally only in a mammalian cell, not in bacteria. Furthermore, in some instances a protein produced as the result of genetic engineering will bear traces of the gene fusions through which it was made. For example it may carry a short sequence of amino acids derived from another protein, either because its gene was fused to the gene for the other protein in order to 'drive' transcription, or because the additional sequence was required to signal protein secretion (Fig. 3.1). Such appendages can often be removed, but not necessarily leaving an entirely natural product.

Thus proteins made by genetic engineering may or may not be identical to their naturally synthesized counterparts. But, even if they are not entirely natural, they may still be efficacious and safe. They just need to be tested for efficacy and safety, like any other new drug.

Possible hazards from the genetically engineered organisms

In the early days of DNA manipulation, some thought that genetically manipulated organisms were risky *per se*. However, in 15 years of genetic engineering in the laboratory there has been no evidence to substantiate this fear. Dangerous organisms could certainly arise from genetic engineering involving pathogens, or genes determining pathogenicity, but there are no examples of pathogenicity arising 'out of the blue' as a result of genetic manipulation. It is certainly incumbent on genetic engineers and regulatory agencies to think through the possible properties of their constructions in order to forestall any predictable dangers. One cannot guard against the absolutely unpredictable, but, to judge from the experience of the last 15 years, we hardly need to fear it.

In the UK, the regulatory body is the Health and Safety Executive, acting through the Advisory Committee on Genetic Manipulation (ACMG). The Committee provides a set of guidelines that permits those planning work involving genetic manipulation to submit their own calculation of the category of hazard and therefore of containment, a calculation that can then be accepted or questioned by the regulators.

In the USA the regulatory body is the National Institutes of Health's Recombinant DNA Advisory Committee (RAC), which uses criteria for categorization similar to those employed by ACMG but uses a more prescriptive mode of operation, listing comprehensively the levels of containment required for each host/vector combination. Most countries have followed the American rather than the British model, probably because it is more in line with their legislative practice. The universally employed principle is case-by-case assessment.

Safety in industrial practice

The regulatory procedures of ACMG and RAC were designed for the laboratory situation, where the amount of material being handled is small, the working area is restricted to a few rooms, and the people involved are relatively few in number and usually highly trained. The application of recombinant DNA technology on the industrial scale has required some new thinking.

A lead was provided by a 1986 study by the Organization for Economic Co-operation and Development (OECD, in effect speaking for developed Western countries) entitled *Recombinant DNA–Safety Considerations* [6]. This study incorporates industrial use of both genetically engineered and non-engineered organisms into the concept of Good Industrial Large-Scale Practice (GILSP). GILSP includes (in the context of genetic manipulation) the following criteria for safety of the engineered organism:

1 the host organism should be non-pathogenic and have a long record of safe industrial use;
2 the vector/insert combination should be free of predictable hazard, and should be as small as possible to minimize the chance of accidental inclusion of genes of unknown function; it should have minimal chance of transfer to other organisms, and in particular should be incapable of transferring antibiotic resistance;
3 the engineered organism should be at least as safe as the natural organism from which it was derived, and without harmful consequences in the environment.

As regards industrial practice, the following points are made:

1 careful attention needs to be given to the fermentation equipment, since this provides the principal physical containment;
2 there must be standardization and rigorous implementation of safe operating procedures, including appropriate training of all employees, both management and workers;
3 detailed attention must be given to those parts of the facility outside the immediate production area, and to work practices therein.

The OECD report pays great attention to human as well as to physical considerations. It emphasizes that the best-designed hardware will not prevent accidents unless all those involved are well trained and aware, and also participate in continued assessment of the adequacy of agreed procedures. To ensure such participation, effective local biological safety committees are seen as essential (just as they are in research laboratories), part of their brief being to monitor adherence to accepted codes of practice. Finally, there is throughout the report a recognition of the need for flexibility, as every case may be expected to have unique features. There will inevitably be differences here from small-scale use of engineered organisms, where procedures can usually be standardized for a given category of hypothetical hazard. In industry there may be several intermediate stages – including pilot plant and start-up – between the laboratory bench and the final production process. These may require a combination of procedures drawn from several different nominal categories.

A question that has not so far become critical is the maintenance of adequate containment procedures in countries where technical skills are less well established. Transfer of advanced fermentation technology to the Third World may not be an immediate prospect, but it will inevitably happen. Every effort should be made to ensure that the necessary training in safety procedures is exported along with the biotechnology and hardware.

Conclusions

Some medically useful products of genetically manipulated organisms are already commercially available. There is little doubt that the number and range of products will grow rapidly.

Work with microorganisms on a small scale, and their industrial employment on a large scale, can involve real hazards of which we already have much experience; as a result, their deployment is already subject to a body of rules and procedures.

We are also already familiar with the hazards of scale-up in the context of chemical and pharmaceutical industry. A chemical or product that needs to be handled with only a moderate degree of care in milligram or test-tube quantities in a laboratory, can become an environmental disaster if released in vast quantities into the River Rhine. Regulations in place in most countries are supposed to rule out such disasters, but experience shows that they are vulnerable to human error.

The advent of genetically engineered organisms in industry seems certain to bring tighter regulation. All work involving laboratory use of such organisms is subject to statutory regulations over and above those applied to microorganisms generally, and these regulations appear to have been adequate. Much thought had gone into the problems posed purely by scale-up, and there seems to be every prospect of their being satisfactorily dealt with.

In chemical industry, many of the substances used on a large scale are clearly highly dangerous, and concern about the maintenance of proper safety standards is fuelled from time to time by more or less disastrous accidents. In the industrial application of genetic engineering, the hazards are hypothetical and, in all probability, there will be no disasters. The most difficult challenge, but one that will nevertheless have to be met, will be to maintain high standards of vigilance even in the absence of practical demonstrations of the consequences of negligence.

References

1 Davies, J. E. (1988) Engineering organisms for use. *The Release of Genetically Engineered Micro-Organisms* (eds M. Sussman, C. H. Collins, F. A. Skinner and D. E. Stewart-Tull), pp. 21–28. Academic Press, London.
2 Baldari, C., Murray, A. H., Ghiara, P., Cesareni, G. and Galeotti, C. L. (1987) A novel leader peptide which allows efficient

secretion of a fragment of human interleukin-1 in *Saccharomyces cerevisiae*. *EMBO J.* **6**: 229–234.

3 Winter, G. P. (1989) Antibody engineering. *Phil. Trans. R. Soc. Lond. B.* **324**: 537–547.

4 Ward, E. S., Gussow, D., Griffiths, A. D., Jones, P. T. and Winter, G. (1989) Binding activities of a repertoire of single immunoglobulin variable domains secreted from *Escherichia coli*. *Nature* **341**: 544–546.

5 Hopwood, D. A. (1989) Antibiotics: opportunities for genetic manipulation. *Phil. Trans. R. Soc. Lond. B* **324**: 549–562.

6 Organization for Economic Co-operation and Development (1986) *Recombinant DNA–Safety Considerations*.

4 | GENETICALLY MANIPULATED MICROORGANISMS FOR USE IN AGRICULTURE

Most large-scale releases of genetically manipulated microorganisms, within which category we may include fungi, bacteria and viruses, are likely to be for agricultural purposes. The use of microbes for the improvement of agricultural processes of various kinds is already big business, as evidenced by the fact that a 280-page report on the marketing of microbial inoculants in Europe, including profiles of 37 of the major manufacturers and suppliers, has recently been offered for sale to those in the business at $3200 per copy.

The objectives will be the improvement of plant or animal nutrition, the control of pests, the prevention of frost damage to crops and, very probably, the processing of animal feeds and the degradation of farm wastes. This chapter reviews the possibilities in each of these areas.

Animal nutrition

Probiotics

The nutrition of farm animals is influenced very significantly by the rich bacterial gut flora that they harbour. It has long been known that low levels of antibiotics, added to animal feed, can improve the efficiency of conversion of feed to liveweight. The effect is probably due to the inhibition of the growth of detrimental bacteria. More recently it has been found that the feeding of live non-pathogenic bacteria of various species can have a similar effect, presumably because the fed bacteria compete with and partially displace the detrimental species. Non-pathogenic bacteria used in this way are called probiotics, and they are already being marketed in large quantities. There is some doubt about the true value of the strains at present in use, but the evidence is sufficiently positive to encourage research into ways of making them more effective. Genetic engineering of probiotics is one obvious avenue of research, though this approach is at present limited by lack of understanding of the internal ecosystem in which probiotic bacteria have their effects.

Rumen bacteria

The nutrition of cattle and sheep is highly dependent on the bacteria of the rumen, which are responsible for the breakdown of ingested herbage material to utilizable carbohydrate. It seems very possible that attempts will be made to modify rumen bacteria so as to broaden the range of plant products that they can deal with. For example, the ability to degrade lignocellulose, ordinarily the least digestible part of plant-cell walls, might be seen as a useful property to introduce into the rumen microflora. Genes encoding the necessary enzymes are known in fungi, and it would certainly be possible to transfer them to rumen bacteria. A likely difficulty with such a strategy would be that of establishing the manipulated bacteria in competition with normal rumen organisms. Indeed, safety considerations might dictate that they should not be competitive, in which case they would presumably have to be constantly replenished by addition to the animal feed.

Silage

The production of silage for animal feed has increased in importance as haymaking has declined. The fermentation required to preserve the nutrients in the grass is traditionally carried out by the microorganisms naturally present, sometimes assisted by the addition of formic or sulphuric acid. But already there is a substantial business in microbial inoculants for improving the efficiency of the fermentation. The bacteria used for inoculation into silos may well be targets for genetic engineering to improve their performance.

Plant nutrition

Root-associated microorganisms

There is already a large market for microorganisms that improve plant nutrition, particularly for root-colonizing bacteria of the genus *Rhizobium* that fix atmospheric nitrogen in the root nodules of leguminous plants such as peas and beans. Many years of research have shown that rhizobia have the potential to satisfy the nitrogen requirement of nodulated legumes, and they are routinely inoculated into the soil when leguminous crops are introduced into areas in which suitable indigenous *Rhizobium* strains are lacking. Much research is going on into methods for improving the nitrogen-fixing capacity of *Rhizobium* strains. Field trials have already been conducted in the USA of a genetically manipulated *Rhizobium* strain, for use with lucerne (*Medicago sativa*), that fixes more nitrogen in the root nodules than the strains currently in use. Interest in *Rhizobium* may extend beyond its role in nitrogen fixation; it is at least possible that it can be modified so as to confer other useful properties on plant roots, such as the capacity to kill disease organisms.

The root nodules of legumes are not the only example of mutually beneficial interaction between soil microorganisms and plants roots. Intimate associations between roots and certain fungi, called mycorrhizas are very common. The fungal partners seem to make soil nutrients more readily available to the plant. Methods for the application of genetic engineering to fungi have been greatly improved over the last five years, but before they can be applied to the improvement of mycorrhizal fungi a much better understanding of the mycorrhizal relationship will be required; at present it is not at all clear which features of the fungus one would wish to alter.

Free-living bacteria

There have been reports that plant nutrition can be improved through the addition to the soil of free-living nitrogen-fixing bacteria (*Azotobacter* species), phosphate-solubilizing bacteria and various cocktails of supposedly useful species. Reproducible results from the use of such inoculants have been hard to come by, but there will probably be further trials, including genetically manipulated bacterial strains, as knowledge of plant–microbe interactions increases.

Control of pests and diseases

Applications of DNA technology to the development of vaccines against animal disease are dealt with in Chapter 6. Here we discuss applications intended to protect crop plants.

Root-associated bacteria

Considerable attention is being given to microorganisms that inhibit other, pathogenic, microorganisms in the soil. One of the best known examples is the bacterium *Agrobacterium radiobacter*, which can be used to control crown gall disease, caused by the related species *Agrobacterium tumefaciens* (the great significance of which for plant genetic engineering is explained in Chapter 5). The non-pathogenic *Agrobacterium* species *A. radiobacter* can control crown gall by producing a toxin, called an agrocin, which acts against *A. tumefaciens* but not against the species that produces it. Both the production of the toxin and resistance to its effects are determined by a plasmid harboured by *A. radiobacter* but not by *A. tumefaciens*. There is a possibility that the plasmid might be transferred to *A. tumefaciens*, which would obviously vitiate the whole plan of control. Approval has recently been granted in Australia for field testing of an *A. radiobacter* strain in which the DNA sequence of the plasmid has been manipulated so as to make it non-transferable.

Other kinds of root-associated bacteria have also attracted attention. For example, a leading American company has field-tested a genetically manipulated derivative of a root-colonizing

Pseudomonas species. The commercial potential of this approach to plant protection remains to be tested. It seems likely to be less predictable and efficient than making the plants themselves resistant, if that can be achieved. Genetic engineering of crop plants is discussed in Chapter 5.

Insect pests

The control of insects was once considered as solely a chemical problem, but it can also be tackled through the use of biological control agents including toxin-producing bacteria and viruses. Genetic engineering promises to make such agents more effective.

Bacillus thuringiensis *toxin*
The bacterium *B. thuringiensis* produces a crystalline protein that is extremely toxic to certain species of moths. Different strains of the bacterium make different forms of the toxin, each with its own range of insect targets. Spraying of the dead bacteria is already a well-tried method of control of moth caterpillars.

The present limitations of *B. thuringiensis* toxin relate to the range of species against which it is active, its low persistence, and the difficulty of producing it in sufficiently large amounts for commercial use. All of these problems can in principle be dealt with by genetic manipulation. For example, the range of susceptible species can be altered by recombining the toxin-encoding sequences from bacterial strains active against different kinds of insects. This is at present being tried by a number of companies.

Persistence of the toxin is mainly limited by its sensitivity to ultraviolet light and could in principle be dealt with by appropriate changes in its structure, which could be achieved by controlled changes in the encoding DNA. Information on the relation between the structure of the protein, its toxicity and its ultraviolet sensitivity could, in the last resort, be obtained by trial and error. Finally, production of the toxin could be improved by putting the gene under the control of a strong promoter and introducing it into a more rapidly growing bacterium, such as *E. coli*.

None of the expedients just outlined would necessarily involve the release of live genetically engineered bacteria into the environment; the toxin could, as before, be applied as a spray of dead bacterial cells. But there is also the possibility of inserting the toxin gene directly into the genome of the plant that one wanted to protect, an alternative that is discussed in Chapter 5.

Insecticidal viruses
Another line of attack against insect pests is through the use of their own pathogenic viruses. Currently, much attention is being given to viruses of the baculovirus group, each member of which causes disease in only a limited range of insects, varying from one virus strain to another. DNA manipulation might both make them even more deadly to the target insects and reduce their ability to persist in the environment. One effective way of achieving the latter objective is to delete the virus gene specifying the protein that normally forms the protective coat around the virus particle. Tests have shown that a virus so modified can still be infectious to moth caterpillars but have a very short survival time on the surfaces of leaves on to which it is sprayed or on the soil beneath [2].

Ways of increasing the lethality of the virus include, most obviously, the insertion into the virus genome of a *B. thuringiensis* toxin gene. At least one American research institute is understood to be working towards this objective.

Control of frost damage

Whether or not a plant is damaged by frost depends to some extent on whether ice crystals form on the surfaces of the leaves. If the temperature falls only a few degrees below freezing point, ice may not form unless there are nuclei around which the crystals can start growing. One important source of nuclei are cells of certain bacterial species, of which *Pseudomonas syringae* is often the most important. It is now known that the propensity of cells of this species to act as centres for ice formation depends on a particular protein on the cell surface. There is an important market for this protein in connection with the making of snow for ski slopes. Its advantage to the bacterium, presumably, is the release of nutrients from the frost-damaged cells. Deletion of the

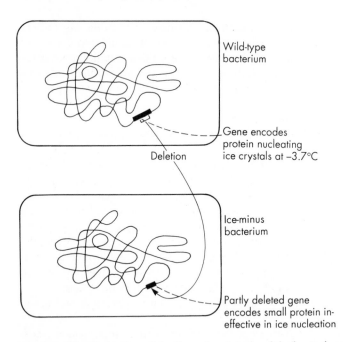

Wild-type bacterium

Gene encodes protein nucleating ice crystals at −3.7°C

Deletion

Ice-minus bacterium

Partly deleted gene encodes small protein in-effective in ice nucleation

Fig. 4.1 The derivation of the 'ice-minus' strain of the bacterium *Pseudomonas syringae*.

gene encoding this protein results in an 'ice-minus' strain of *Ps. syringae* (Fig. 4.1). Large numbers of ice-minus bacteria sprayed on to the leaves of the crop that one wishes to protect will compete with and effectively exclude the smaller numbers of 'ice-plus' naturally present. Thus ice formation should be avoided, at least down to temperatures of about −10°C. In California this scenario has now been checked out in practice and found to work. The procedure has obvious potential in connection with fruit and vegetable farming – oranges and strawberries are two of the most obvious candidates for protection.

This example involves only the most modest kind of genetic manipulation. The bacterial genome has nothing added to it and has just been subjected to a deletion of a kind that could, and doubtless does, occur spontaneously at a low frequency in nature.

If this deletion was conducive to invasiveness of the bacterium it could already have become established in the wild by natural selection. In fact, pilot experiments indicate that the ice-minus *Pseudomonas* does not persist in the environment in competition with the natural bacterial flora [3].

Disposal of farm wastes

Farm wastes are becoming a serious problem because of the intensification of agricultural production. With the decline of mixed farming, there is no longer the outlet that there once was for straw for farmyard manure. Intensive rearing of livestock in indoor units generates large amounts of animal waste that may overload sewers or pollute streams and rivers. Effluent from silos adds to the problem.

Animal waste and silage effluent can be processed in slurry tanks by appropriate microorganisms, some of which might be susceptible to improvement through genetic engineering. Surplus straw is at present largely disposed of by burning, which creates a public nuisance and is likely to be increasingly restricted. The obvious alternative of chopping it and ploughing into the soil has its own serious drawbacks: possible adverse effects on the germination of the following crop and the nourishment of plagues of slugs. Procedures designed to hasten the breakdown of straw through the use of 'cocktails' of fungi and bacteria, selected for their ability to digest cellulose and fix nitrogen, have been developed in a number of countries, usually with little success. Here again there is the possibility of increasing the effectiveness of straw degrading organisms through genetic engineering.

Possible hazards – problems of monitoring

Bacteria

There is no reason to suppose that any of the kinds of genetically manipulated bacteria discussed in this chapter would pose a direct threat to human or animal health. The concern is rather that if they were too robust and competitive they might establish themselves in place of other, natural microbial species, perhaps with complex effects on wild life and soil fertility that would be difficult to predict. There are no precise hypotheses of danger that could be tested, but natural bacterial populations are never so completely described as to make it possible to assert that all possible dangers are forseeable.

While it might well be possible to establish, with reasonable assurance, the safety of the genetically manipulated organism that was initially released, it is a far more difficult matter to identify all the possible transfers of its genetically manipulated DNA to other species, and to predict the effects of such transfers. We know of many kinds of plasmids and bacteriophages that can act as vectors for cross-specific transfer of DNA, and, while it would be easy to make sure that an engineered organism was not released with such a vector already in it, it would be impossible to rule out its later acquisition of a phage or plasmid from the wild.

Our knowledge or plasmid natural history is certainly far from complete. Recently, a plasmid has been identified that will transfer DNA between the relatively distantly related 'Gram-positive' and 'Gram-negative' bacterial classes. Still more surprising is a very recent report [4] of plasmid-mediated transfer between a bacterium and yeast. What we already know about the promiscuity of DNA transfer in the microbial world suggests that, if we want to be reasonably sure that artificially manipulated DNA will not persist in the environment, we had better make sure that the bacterium in which it is introduced has itself such a limited capacity for survival that it has only a minimal chance of donating its DNA to other organisms.

Arranging for limited survival without jeopardizing the whole object of the exercise is a difficult problem. To be any use, the novel microorganism has to persist long enough to do the job it was designed for – colonize roots, exclude its ice-minus progenitor, degrade the straw, or whatever. So it is necessary to design for short-term survival and medium-term disappearance. The general way one would hope to achieve this would be to build into the new genotype some handicap that was not immediately crippling but imposed a disadvantage as compared with competitive wild organisms. The short-term survival of the engineered microorganism could be ensured by conferring on it some immediate benefit, which would usually be large inoculum size and perhaps some especially favourable but temporary habitat, such as an appropriate crop to interact with or a mass of straw to digest. When this special benefit lapsed natural selection could be relied upon to take its course.

However reasonable the plan, it would obviously need to be tested empirically before the engineered organism could be approved for release. Realistic testing short of release is obviously more difficult to achieve with microorganisms than with plants that are large enough to see with the unaided eye. A microorganism is effectively out of sight and out of control from the moment it is inoculated into an agricultural field. So preliminary testing has to be conducted in so-called microcosms – contained laboratory environments that simulate field conditions as closely as possible [5]. If such tests indicate that the novel organism does not long survive in competition with natural organisms then it may be reasonable to proceed with small scale release into open fields, with monitoring of the subsequent changes in the microbial population.

Monitoring is itself a problem. In order to count the number of surviving engineered bacteria in a given volume of soil it is necessary to endow them with some specific characteristic, or marker, than can make them countable even in the presence of large numbers of other microorganisms. One obvious type of marker is antibiotic resistance, which can be easily introduced into the cells that one wishes to monitor. But the deliberate release into the environment of antibiotic-resistant bacteria is generally regarded as highly undesirable, for fairly obvious reasons. Fortunately, other kinds of marker are available that pose no arguable risk, for example the ability of the bacterium to use the milk sugar lactose [6]. A general limitation, however, of all markers that depend on metabolic activity for their expression is that some bacteria can enter a cryptic, dormant phase during which they display no activity that could make them visible [7].

A recently developed technique, the polymerase chain reaction (PCR), enables the special DNA of any cell to be detected, whether the cell is metabolically active or not. This reaction, using repetitive priming of DNA synthesis, enables any known DNA sequence to be indefinitely amplified from a tiny amount of starting material (see Fig. 6.3, p. 78). A paradoxical drawback of this technique is its exquisite sensitivity; it is difficult to infer from the strength of the amplified signal how much DNA was originally present, and a false-positive result could follow from chance contamination with even a few molecules of the DNA sequence being monitored.

Viruses

The insect viruses proposed for genetic manipulation and use as insecticides each infect only a limited range of insect species and are no direct hazard to people. However, one would want to be sure that the range of insects that they attacked did not extend to beneficial insects such as bees or ladybirds, or aesthetically pleasing and harmless ones such as most butterflies. The baculo-viruses on which current efforts are concentrated do indeed have great species specificity and no damage to innocent insect by-standers would be expected. Manipulation to reduce the persist-ence of the virus, as described above, would provide a second line of defence against unwanted effects.

The host specificity of an engineered virus can be easily checked against a range of caterpillars of different species. Moni-toring of virus persistence in the soil or on vegetation is also relatively straightforward through infectivity tests on the target species.

How much should we fear interspecific DNA transfer?
Evolutionary considerations

The possibility of transfer of manipulated DNA to species for which it was not intended needs to be considered in connection with most forms of genetic engineering, but it is particularly relevant in the context of this chapter. The degree of uncertainty about what is going on in the largely hidden world of microbial ecology is such that, if one was really convinced that combining DNA from different species was likely to lead to new kinds of invasive organisms, it might well be better to desist altogether from the genetic engineering of bacteria.

Paradoxically perhaps, some of the considerations that make unexpected genetic consequences of release of engineered bac-teria so difficult to rule out also provide some grounds for reassurance about possible hazard.

We know that bacteria have several different natural ways of exchanging DNA between different species – sometimes, indeed, between species that are only very remotely related. The least specific mode of DNA transfer of all is transformation by free DNA, which many bacterial species can pick up from the environment with various degrees of efficiency. DNA acquired in this way does not even have to be bacterial in origin; we know of no reason why it could not come from any kind of dead plant or animal with whose decaying remains the bacterium came into contact. We must therefore assume that interspecific DNA transfer is going on in bacteria all the time, no doubt at a very low frequency per cell but involving large numbers of bacteria in the natural world as a whole. If this were an important factor in the evolution of new successful organisms one would expect to find evidence of it in the many comparisons that are now being made between the DNA sequences of diverse species [8].

So far the evidence for the presence in wild species of tracts of DNA acquired from other organisms is meagre. DNA sequences found in particular species bear the hallmarks of the branch of the evolutionary tree to which the species are considered to belong, without obvious intrusions from other lineages. The fact that it is possible to construct consistent evolutionary trees from DNA sequence data, and that the trees come out very much the same whichever gene is chosen for the exercise, is evidence against frequent 'lateral' transfer of DNA between unrelated species, which, if it happened, could thoroughly confuse the picture.

Nevertheless, transfer of genes between widely different organisms may occur. Several of the most convincing examples come from the genus *Streptomyces* (famous as a source of antibiotics). One species produces a protein that looks remarkably like calmodulin, which, in eukaryotic organisms mediate the response of cells to calcium, but is otherwise unknown in bacteria. Several species have glutamine synthetase (an enzyme of central importance in nitrogen assimilation) of a kind that is characteristic of plants rather than of bacteria. Both kinds of glutamine synthetase are found in *Rhizobium*, suggesting, perhaps, that this genus has captured the plant form of the enzyme without giving up the bacterial form.

The examples just mentioned suggest gene transfer to bacteria from plants (or possibly animals, in the case of calmodulin). There is one well-known example of possible transfer from animal to plant. The root nodules of leguminous plants contain an oxygen-binding protein (leghaemoglobin) encoded by a gene with some remarkable similarities to the genes for mammalian haemoglobins.

We still have insufficient information to assess the frequency of effective gene transfer between very dissimilar organisms, but the rather small number of convincing examples suggest that it is very low.

If, in spite of the ample opportunities that appear to exist for interspecific DNA transfer in the microbial world, there really has been very little persistence of DNA transferred between widely dissimilar bacteria, one might tentatively conclude that genes, naturally selected for optimal function in one kind of organism, will be very unlikely to fit easily into the well-adapted and integrated system of another. Such reasoning would suggest that cells carrying exotic genes tend to be eliminated by natural selection. If this has been true on the evolutionary time scale for foreign DNA acquired naturally, which in principle could be of any kind whatsoever, it seems likely to apply to foreign DNA inserted by genetic engineering.

Some would argue that the natural world – a huge unregulated constantly on-going experiment – can in time come up with virtually anything that could be fabricated in the laboratory. For the present, however, it may be as well to heed the more cautious verdict of a group of American ecologists [9]:

> The available scientific evidence indicates that lateral transfer among micro-organisms in nature is neither so rare that we can ignore its occurrence, nor so common that we can assume that barriers crossed by modern biotechnology are comparable to those constantly crossed in nature.

References

1 Sussman, M., Collins, C. H., Skinner, F. A. and Steward-Tull, D. E. (eds) (1988). *The Release of Genetically-Engineered Micro-Organisms*, 302 pp. Academic Press, London.
2 Bishop, D. H. L., Entwistle, P. F., Cameron, I. R., Allen, C. J.

and Possee, R. D. (1988) Field trials of genetically engineered *Baculovirus* insecticides. *The Release of Genetically-Engineered Micro-Organisms* (eds M. Sussman, C. H. Collins, F. A. Skinner and D. E. Stewart-Tull), pp. 143–179. Academic Press, London.

3 Lindow, S. E. and Panopoulos, N. J. (1988). Field tests of recombinant ice-minus *Pseudomonas syringae* for biological frost control in potato. *The Release of Genetically-Engineered Micro-Organisms* (eds M. Sussman, C. H. Collins, F. A. Skinner and D. E. Stewart-Tull), pp. 121–138. Academic Press, London.

4 Heinemann, J. A. and Sprague, G. F. Jr. (1989) Bacterial conjugative plasmids mobilize DNA transfer between bacteria and yeast. *Nature* **340**: 205–209.

5 Lindow, S. E., Panopoulos, N. J. and Fraley, R. T. (1989) Genetic engineering of bacteria from managed and natural habitats. *Science* **244**: 1300–1307.

6 Drahos, D. J. *et al.* (1988) Pre-release testing procedures: field test of a *lacZY*-engineered soil bacterium. *The Release of Genetically-Engineered Micro-Organisms* (eds M. Sussman, C. H. Collins, F. A. Skinner and D. E. Stewart-Tull), pp. 193–206. Academic Press, London.

7 Colwell, R. R., Brayton, P. R., Grimes, D. J., Roszak, D. B., Huq, S. A. and Palmer, L. M. (1985). Viable but non-culturable *Vibrio cholerae* and related pathogens in the environment: implications for release of genetically engineered microorganisms. *Bio/Technology* **3**: 817–820.

8 Ambler, R. P. (1985) Protein sequencing and taxonomy. *Computer-assisted Bacterial Systematics* (eds M. Goodfellow, D. Jones and F. G. Priest), pp. 307–335. Academic Press, London.

9 Tiedje, J. M., Colwell, R. K., Grossman, Y. L., Hodson, R. E., Lenski, R. E., Mack, R. N. and Regal, P. J. (1989) The planned introduction of genetically engineered organisms: ecological considerations and recommendations. *Ecology* **70**: 298–315.

5 | GENETIC ENGINEERING OF AGRICULTURAL PLANTS

Plant cell manipulation and regeneration

The application of DNA technology to the genetic modification of plants depends substantially on techniques for the culture of plant cells, originally taken from stem or leaf, as undifferentiated callus tissue on artificial nutrient medium supplemented with appropriate plant hormones. Though, as we see below, techniques for transfer of DNA to whole plants are now becoming available, plant genetic engineering up to now has depended on transformation of cells in culture.

Transforming cell cultures is, however, no use unless they can be made to regenerate into roots, shoots and eventually whole plants. And while cell cultures can be obtained from nearly all plants, regeneration can be achieved easily in only a few species, mostly members of the potato family (Solanaceae), including tobacco, tomato and petunia. With somewhat more difficulty, regeneration can be achieved in members of the cabbage family, which includes a number of important vegetables (cabbage, sprouts, radish, turnip) as well as mustard and rape. Even more difficult are the grasses, including the vital cereal crops rice, wheat, barley, rye, oats and maize, and legumes such as peas and beans. This is at present a serious limitation of plant genetic engineering based on cell culture.

Techniques for introducing DNA into plant cells

The Agrobacterium T-DNA system

Transfer of foreign DNA into plant cells, as into bacteria, can be greatly facilitated by the use of a vector, which may be either a plasmid or a virus. By far the most successful plant vector is the plasmid (Ti) harboured by the soil bacterium *Agrobacterium tumefaciens* [1,2]. This bacterium can parasitize and provoke tumour formation in a wide range of broad-leaved plants, but not in grasses and related cereal crops. The tumours can be excised from the infected plants and grown in culture, retaining their tumorous character even when the bacteria that provoked the tumour are eliminated. The explanation for this permanent transformation of cell type is that *A. tumefaciens* harbours a DNA plasmid with the unique property of being able to inject a portion of itself – the T-DNA segment – into any plant cell with which the bacterium is in contact.

Once it has entered the cell the T-DNA becomes integrated into one or more of the chromosomes of the cell nucleus, and is thereafter replicated along with the host cell chromosomal DNA. It contains genes determining the production of a plant hormone that stimulates growth of the tumour. Other T-DNA genes encode enzymes for the synthesis of certain nitrogenous compounds, opines, that are especially good nutrients for the bacterium; this seems to be the point of the tumours from the bacterium's point of view.

Although T-DNA is transmitted by infection it is not infectious in its own right. It does not carry the genes necessary for its own transfer – these are present elsewhere in the plasmid and are left behind in the bacterial cell. The transferred T-DNA is like a spent bullet separated from its cartridge.

The Ti plasmid can be isolated as free DNA, and its T-DNA segment modified by the standard procedures of cutting and ligation outlined in Chapter 2 (p. 18). The tumour-promoting genes can be excised and a gene from another source inserted instead. After reintroduction into *Agrobacterium* the plasmid can transfer its modified T-DNA into cultured plant cells that can then, with varying degrees of difficulty, be regenerated into a whole plant every cell of which carries the exotic gene (Fig. 5.1).

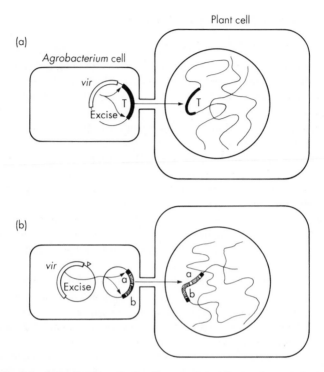

Fig. 5.1   Introduction of a foreign gene into tobacco (or potato or tomato) by means of the *Agrobacterium tumefaciens* Ti plasmid. (**a**) In normal *Agrobacterium* infection the T-DNA segment of the Ti plasmid is excised and transferred into the plant genome by the action of virulence (*vir*) genes present in Ti. The T-DNA induces tumorous growth in the recipient cells. (**b**) Represents an *Agrobacterium* strain with two plasmids, one a Ti plasmid that has its T-DNA deleted but its *vir* genes intact, and the other a plasmid carrying two alien genes, (a) conferring antibiotic resistance and (b) coding for a function that one wishes to introduce into the plant; these two genes are flanked by T-DNA termini (black segments). The *vir* functions of the T-deleted Ti plasmid act on the T-termini of the second plasmid and transfer the alien genes into the plant cell, where they become integrated into the genome. Plant cells that have acquired the engineered T-DNA are selected as resistant to the antibiotic and regenerated into whole plants, which should also carry the desired (b) gene.

It is necessary to have some means of identifying and selecting clones of cultured cells that have picked up the T-DNA and its passenger gene. With normal T-DNA this poses no problem since the formation of the tumour is obvious. But disarmed T-DNA vectors with their tumour-producing genes removed do not signal their presence unless they are endowed with some other genetic marker. Resistance to the antibiotic kanamycin, determined by a bacterial gene, has often been used and has the advantage of being automatically selectable if kanamycin, or a related drug, is added to the culture medium. An alternative is to use a marker that can be detected visually such as a gene encoding ß-glucuronidase, an enzyme that can be used to generate a blue pigment (Fig. 5.2).

Even though it does not induce tumours in cereal crops, the Ti plasmid can still inject its T-DNA at least into maize, and presumably into other grass relatives as well. This was shown by modifying the T-DNA so that it carried into the maize plant a virus infection that could not otherwise have gained access to the plant without its normal insect vector. This process, called 'agroinfection', might have some potential for genetic engineering if, instead of the virus, the T-DNA carried another gene or set of genes that one wanted to get into the plant. However, although T-DNA can get into maize cells, there is as yet no evidence of its integration into maize chromosomes, as would be necessary for stable transformation. This possible method for transforming cereals may, in any case, be made redundant by methods of transformation without vectors (see p. 52).

The use of viruses as vectors

The possibility of using plant viruses for plant genetic engineering has received much attention [2,3]. The general plan is to manipulate the virus genome to remove the genes responsible for disease symptoms and for propagation of infective virus, while retaining those required for initial infection and replication. The excised genes can, in principle, be replaced by DNA sequences designed to transform the plant in some desired way. In order to get stable transmission to following generations the transforming DNA needs to be integrated into the chromosomes.

The most promising plant virus for use as a vector is cauliflower mosaic virus (CaMV), which infects plants of the cabbage family and, less efficiently, a number of other species. Like most other plant viruses it has RNA rather than DNA in its infective particles, but, unusually for plants, it is a retrovirus, a kind of virus that is more widespread and more extensively studied in animals (see p. 74). The RNA is not replicated as such after infection but is copied into DNA by the enzyme reverse transcriptase, encoded by the virus. The DNA copy of the viral RNA is then replicated in the plant cell and transcribed to make more infectious RNA. This way of life is characteristic of the retroviruses. The fact that they bring their own efficient mechanisms for replicating as DNA in the host cell makes them especially suitable as vectors.

In spite of its apparent suitability, CaMV has not so far proved very easy to adapt for use as a vector. It is not very tolerant of modifications to its genome and may not be able to accept foreign DNA segments of useful size without impairment of its ability to replicate. The problem of 'disarming' the virus so that, even though able to infect and replicate its DNA, it can no longer form infective particles or cause disease, has also been difficult to solve.

Vectorless transformation

The feasibility of DNA transfer without a vector depends upon the curious fact, true of both plant and animal cells, that artificially introduced DNA of any kind tends to become integrated even if it is quite unrelated to any DNA normally present. Such non-homologous integration occurs into the chromosomes more or less at random.

The first introductions of vectorless DNA into plant cells depended on the use of protoplasts – cells with their outer walls digested away and separated from the surrounding medium only by the thin cell membrane. Uptake of DNA across the membrane can be facilitated by pulses of high-voltage electricity, a procedure called 'electroporation'. Another way is injection by means of a micro-syringe. The most recent method is to literally 'shot-gun' the DNA into intact cells, either in callus culture or in intact embryos or shoot apices. Minute tungsten pellets are coated with DNA and fired into the cells by means of a small explosive cartridge. The punctures made by the pellets are too fine to cause serious damage and a proportion of them, with their cargo of DNA, lodge in the cell nuclei to initiate clones of transformed cells. This apparently unlikely method actually works (Fig. 5.2).

Fig. 5.2   The effect of shooting DNA containing a visible marker gene (for the enzyme ß-glucuronidase) into a plant callus culture. The culture medium contains a compound that is turned blue by the enzyme. (Photograph by courtesy of Professor R. B. Flavell.)

Obtaining the desired expression of the introduced genes

If a gene artificially introduced in a DNA fragment is to have its desired effect it must be 'switched on' at the right levels and at appropriate times and places in the development of the plant. This means that it must be flanked by appropriate transcription-controlling sequences – promoters and enhancers – that will respond to signals generated from elsewhere in the genome. Because integration is essentially at random, a newly integrated gene is very unlikely to find appropriate controlling sequences at its site of insertion. It is therefore necessary to provide such sequences in the DNA construction that is used to transform the plant. This is becoming increasingly possible as more and more specific control sequences, determining, for example, gene activity in one tissue rather than another, or in response to specific stimuli such as light or wounding, are identified and cloned.

A potential problem arising from random integration is the possibility that the rather large pieces of DNA that it is necessary to use could interfere with the expression of other genes nearby. Such effects seem not to be very common in plants, which tend to have large chromosome tracts into which DNA can be inserted without disrupting anything else. But they do occasionally happen.

Limitations due to recessiveness of some desired genes

In yeast, the most amenable microorganism, transformation with an artificially manipulated gene commonly occurs through its insertion at the site of its normal counterpart, which is consequently replaced or disrupted. In higher plants and animals such targeted integration is a much rarer event, and transforming DNA is usually integrated at some apparently random locus, leaving the normal gene copy untouched. So transformation of the phenotype depends on the effect of the introduced gene being dominant to the normal gene function, or alternatively of a kind not normally present in the plant at all. This is a serious limitation. Some of the most useful gene variants are recessive or at best semi-dominant, like the dwarfing genes of wheat and rice that made a big impact (the 'Green Revolution') on agricultural productivity in some Third World countries in the 1960s and 1970s. If, in order to exert its effect, a new gene variant has to replace its normal counterpart, the only way may be selection of a rare random mutation in the intact plant rather than by direct DNA manipulation. However, a gene that is virtually recessive in single copy can sometimes impose its effect if introduced in multiple copies, which can be done. And, fortunately, many desirable genes are of dominant effect even in single copy.

What kinds of genes may plant breeders wish to transfer?

Genes for disease resistance

Resistance to disease – for example to mildew and rust in cereal crops – has always been a major objective of plant breeders. They have distinguished two kinds of disease resistance, the first depending on a single gene in each case and the second on the cumulative action of an indefinite number of genes each with a small effect. Single gene resistance is obviously the more amenable to genetic engineering, and much effort is currently being directed towards the cloning of some of the responsible genes, as much with a view to increasing fundamental knowledge as in the hope of immediate practical benefit. The introduction into crop varieties of single dominant resistance genes, often by crossing with naturally resistant wild relatives, is a path already well trodden by plant breeders with considerable success. Unfortunately the method has the serious general limitation that a single resistance gene generally confers resistance only to a single strain of the disease fungus, which can frequently overcome the resistance by a single gene mutation of its own. Such breakdown of single-gene resistance has occurred repeatedly in the history of plant breeding. Resistance based on multiple genes tends to be more durable but is not readily available by genetic engineering because the relevant genes are difficult to identify and clone.

There are interesting possibilities for the development of resistance to viruses [4]. DNA sequences related to genes of the virus, especially those encoding proteins of the virus coat, sometimes confer a useful degree of immunity when integrated and expressed in the plant genome. How this works is not well understood; it may be that the replication of the virus is blocked by a large excess of packaging protein.

Resistance to insect pests

As mentioned in Chapter 4, the bacterium *Bacillus thuringiensis* produces a very potent protein toxin that is active against the larvae of certain butterflies and moths. Dried preparations of the bacterium are already widely used, especially to protect forestry plantations against moth caterpillars. The *B. thuringiensis* gene that encodes the toxin can be introduced into plants under the control of promoter sequences that ensure its expression in leaves. The toxin appears to be entirely harmless to organisms outside a limited range of insect species. Different forms of it, made by different strains of the bacterium, are selective within this range. This kind of crop protection should harm only those species that feed on the crop in question.

Substances that are not usually regarded as toxins may also repel insects. Lectin, an abundant seed protein of legumes is the cause of favism, an adverse reaction to beans, but only people with a particular heritable metabolic deficiency are susceptible. Potatoes with a pea lectin gene expressed in the leaves seem to have a useful degree of immunity to aphid attack. Another plant protein, cowpea trypsin inhibitor, also confers insect resistance on plants which have been engineered to produce it in the leaves.

Herbicide resistance

Now that agriculture in developed countries depends so heavily on chemical control of weeds, it is an obvious move to produce crop varieties that are immune to the herbicide the farmer wishes to use.

One example concerns the herbicide glyphosate, which has its toxic effect through interfering with an essential step in the synthesis within the plant of certain amino acids, essential components of all proteins. The gene coding for the enzyme responsible for this step can be cloned from bacteria and inserted into the plant to boost its ability to maintain amino acid synthesis in spite of the herbicide. Genes conferring resistance to several other kinds of herbicide are now available from various sources and several genetic engineering projects using such genes appear to be close to commercial success [5].

Improved quality of the harvested crop

The most obvious candidates for improvement by genetic engineering are plants grown for their edible seeds – grains and legumes. Proteins, stored in the seeds for the nourishment of the plant embryo during and after germination, are an important component of the harvest. These proteins are the products of a relatively few genes present in multiple copies and active specifically in the endosperm (in cereals) or in the fleshy seedling leaf or cotyledon (in the legumes). Exotic genes, coupled to endosperm or cotyledon-specific promoter sequences, will themselves be expressed in these tissues if introduced into the plant. If introduced in multiple copies, they could have an important impact on the protein composition of the harvest. If the genes were well chosen the effect might be an improvement in nutritional value or some other desirable characteristic such as baking quality of flour.

Extra genes introduced for this sort of purpose need not necessarily be from alien sources; the same general method could be used to alter the balance of a species' own seed proteins. The principle could be extended to other parts of the plant provided that a single gene product made a significant contribution to its overall composition.

Nitrogen fixation

When plant genetic engineering was first discussed, the most widely canvassed goal was the creation of new kinds of crop plants that could obtain their nitrogen from the atmosphere rather than having to be fed nitrogenous fertilizer. Of currently important crops, only the legumes – beans, peas, soybeans and herbage legumes such as clover and lucerne – have this nitrogen fixing ability, and they owe it to the presence in root nodules of bacteria of the genus *Rhizobium*. Nitrogen fixation is catalysed by a complex enzyme, nitrogenase, that is produced by the bacterium. But the conditions for its activity are provided by the root nodule, which is the product of bacterium–plant interaction. Several plant genes may be necessary for nodule formation and function. One that has been characterized encodes leghaemoglobin, a haemoglobin-like pigment that combines with oxygen and is needed for the oxygen-depleted environment within the nodule without which nitrogen fixation cannot occur.

Other bacteria, notably members of the genus *Azotobacter*, also fix nitrogen, but live freely in the soil rather than inhabiting root nodules. They also have been considered for genetic engineering, either to improve their own nitrogen fixing capacity or as a source of nitrogenase genes for incorporation into plants.

There are two reasons for doubt about the feasibility of extending nitrogen fixation to plants in which it has not existed before. One concerns the energy cost to the plant. Nitrogen fixation consumes considerable energy at the expense of carbohydrate (obtained from photosynthesis) that could otherwise be used to fuel plant growth. Legumes evidently balance their energy budget satisfactorily, but they have had a long period of evolution to refine their system. Non-legumes might well require some complex physiological restructuring before they could profit from nitrogen fixation.

The second doubt concerns the feasibility of reconstructing in wheat, or whatever crop was chosen, a structure equivalent to the legume root nodule in providing suitable conditions for nitrogenase to work. It would be quite surprising if the assembly of legume genes concerned with nodule formation were to function in the same way if introduced into wheat. Many features of their new plant host might have to be adjusted in order to accommodate them.

Interest in the possible creation of new nitrogen-fixing species persists but not as a high priority. It would probably be more realistic to set out to improve existing nitrogen-fixing plants.

Elimination of particular gene functions – antisense RNA

In particular cases, DNA manipulation may be used not to introduce new gene functions into plants but to eliminate or weaken functions previously present. An ingenious general

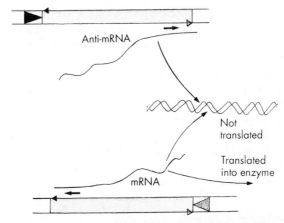

Fig. 5.3   The principle of the use of antisense RNA to silence gene activity. The gene (stippled bar) is cloned, separated from its normal promoter (large stippled arrowhead), and ligated to a strong promoter sequence at the wrong end of the gene (large black arrowhead) to direct transcription of the antisense DNA strand (small open arrowhead) instead of the sense strand (small black arrowhead). The DNA construction is introduced into the plant by cell transformation, preferably in multiple copies. The antisense transcript (anti-mRNA) anneals to the normal mRNA to form a non-translatable RNA duplex.

method depends on inducing the plant to produce large amounts of antisense RNA capable of trapping the messenger in an untranslatable RNA duplex. Antisense RNA is the product of transcription from the wrong DNA strand, as will happen if the gene has its promoter at the wrong end. Figure 5.3 illustrates the principle.

Attempts to achieve gene knock-out by antisense RNA have generally encountered difficulties, but there appears to have been success in one case. In tomatoes an enzyme, polygalacturonase, digests a component of the plant cell wall. The gene encoding this enzyme has been cloned, separated from its normal upstream promoter, and joined, in inverted orientation, to a strong promoter from the cauliflower mosaic virus genome [6]. This construction, introduced into tomato plants by the *Agrobacterium* method, greatly reduced the level of polygalacturonase in the fruit. The practical objective of this work was to produce a tomato variety in which the softening of the skin that normally accompanies fruit ripening is delayed, with obvious potential benefit to marketing.

There may well be other cases in which a crop could be improved by partial or complete knockout of a particular gene function. In principle, any desired loss of gene function ought to be obtainable by selection of an inactivating mutation in the gene concerned without any DNA manipulation at all. The hunt for any particular kind of random mutation is likely, however, to be a long and frustrating exercise. The antisense principle offers a possible short cut to the same end.

Converting plants to entirely new uses

As we shall see in Chapter 6, one of the main current objectives of genetic engineering of animals has been to endow them with the capacity to synthesize pharmaceutically valuable products in their milk. Analogous possibilities with plants have been much less talked about, but they evidently exist.

As a first example of what can be done, a recent paper [7] described the introduction into tobacco plants, by the *Agrobacterium* T-DNA system, of genes encoding mammalian 'heavy' or 'light' antibody protein chains. Under the control of a suitable plant promoter, these alien genes did indeed produce the encoded proteins in the leaves of the plant. When light-chain producing and heavy-chain producing plants were crossed together, the hybrid progeny, carrying both genes, produced complete antibody molecules with the characteristic complex of heavy and light chains. These plant-produced antibodies were shown to be closely similar in several parameters to their animal equivalents. They differed from natural animal antibodies in their specific carbohydrate components, and the extent to which this would militate against their therapeutic usefulness is not yet clear.

The attractiveness of plants as sources of pharmaceuticals lies in the simplicity and cheapness of their culture as compared with industrial microorganisms or farm animals.

Extending the range of plant hybridization

A great many new plant species have arisen in nature by fortuitous doubling of the chromosome number in an initially sterile species hybrid. Such doubled hybrids are called 'amphidiploids' because they have two diploid chromosome sets, one from each parent species. They tend to be fertile, however dissimilar the parent chromosomes sets are from each other, because each chromosome has a like partner that it can pair with and segregate from at meiosis (the process whereby the chromosome number is halved prior to germ-cell formation). They are, in effect, new instant species because, from the outset, any hybrids from crossing back to the diploid parent species are sterile. A comparatively recent example is the grass of salt marshes, *Spartina townsendii*, which was first recorded in 1878 in Southampton Water on the English Channel, and is now established on mud flats in numerous locations round the English and Irish coasts. It apparently arose through chromosome doubling in a hybrid between the native species *S. maritima* and an immigrant from America, *S. stricta*. *S. townsendii* has turned out to be a very vigorous plant, useful in stabilizing mud flats but sometimes a nuisance in clogging waterways. An example of the deliberate use of wide crossing and chromosome doubling is *Triticale*, a wheat–rye compound with some potential as a new crop.

An alternative way of obtaining new amphidiploids is by cell fusion. Protoplasts obtained from cells of different species can be induced to fuse together, and fusion of the cell nuclei commonly follows. If whole plants can be regenerated from the cell culture, the result may be a new hybrid that has never existed before. Because such hybrids arise from the fusion of diploid somatic cells rather than haploid germ cells they are amphidiploids and, potentially, new fertile species. An example was the potato–tomato hybrid that was made some years ago. Unfortunately, it turned out to have neither edible fruits nor useful tubers.

The use of artificial means for obtaining hybrids that could not arise naturally is, in fact, far from new. Many interspecific hybrids fail because of incompatibility between the hybrid embryo and the predominantly maternal constitution of the endosperm, the nutritive tissue of the seed. Such embryos can sometimes be rescued and grown into plants by excising them from the seed and culturing them on artificial nutrient medium. Embryo culture has been used, for example, to produce several novel *Brassica* hybrids.

The whole-cell or whole-embryo techniques discussed in this section are not genetic engineering in the usual sense of DNA manipulation, but they do provide possibilities for producing new kinds of plants that could not be produced by conventional breeding. They may arouse some of the same concerns as have been expressed in connection with genetic engineering proper.

What are the possible hazards?

Possible hazards of the engineered plant itself

Possible new toxins and viruses
Some deliberately conferred novel traits might be seen as potentially obnoxious in themselves, for example as toxins or infectious agents.

For example, the construction of a plant producing its own insecticide in the form of a *Bacillus thuringiensis* toxin might seem to pose environmental concerns similar to those aroused by chemical spraying. In fact, this particular example is one in which the genetic engineering solution should be less controversial than the chemical one. *Bacillus thuringiensis* toxin, because of its great specificity, would be much more precisely targeted against the pest and less likely to be harmful to humans or to wild life that one wished to preserve. It is also more rapidly degraded than most chemicals. However, these reassurances would not necessarily apply to all other toxins that might conceivably be introduced into plants.

Of the few existing modes of plant aggression against humans, hay fever almost certainly causes the most distress overall. Natural molecules occurring on the surfaces of certain kinds of pollen grain provoke quite severe allergic reactions in many people. If new kinds of molecules are to be made in plants as a result of genetic engineering, it will be necessary to be quite certain that they are not allergenic and produced in the pollen.

A very hypothetical risk is that an introduced exotic gene coding for a novel enzyme – an example might be bacterial glucuronidase, which is quite commonly used as a marker in plant transformation (see Fig. 5.2) – might convert a normal plant constituent to a toxin, or perhaps a chemical spray to something more noxious. There is no reason from present biochemical knowledge for expecting such effects, but they cannot be discounted without empirical test.

One of the possible applications of genetic engineering to cereal breeding is the alteration of the protein composition of the grain. It must be kept in mind that a protein that would be a beneficial addition to most people's diet might be harmful to a minority. The natural wheat protein gluten, for example, is severely upsetting to people of a particular genetic constitution.

Schemes for conferring virus resistance by introducing into the plant DNA constructions including virus genes pose more complex problems. It is not impossible that some of these could, by interaction with wild viruses, occasionally give rise to new varieties of virus, perhaps with greater virulence or altered host range. In general, this seems an improbable scenario, but its plausibility will vary depending on what is known about the function of the introduced viral gene and it will have to be assessed on a case-by-case basis.

Possibly enhanced weediness
A rather different cause of concern is the possibility that a plant with a novel engineered trait might turn out to be invasive and become a serious nuisance as a weed. Some of the traits under consideration evidently pose no threat of this kind. There is, for example, no conceivable reason why a change in wheat grain composition intended to improve bread-making quality, or a change in plant conformation to make harvesting easier, should increase the fitness of an escaped seedling in competition with wild plants. But certain improvements in crops might be beneficial to a plant in the wild also.

The most obvious such improvement is disease resistance, but there are reasons for doubting that this in itself could make a crop invasive. By interspecies crosses and selection, plant breeders have been introducing disease resistance into crops for very many years, without creating any new weeds. Disease resistance is, in any case, likely to be less critical for plants in the wild than it is for crops; the latter are unnaturally vulnerable to cross-infection because of their growth in pure stands.

Some of the more speculative plant engineering projects that have been discussed suggest correspondingly speculative hazards. For example, a cereal crop endowed with the ability to fix nitrogen could, just conceivably, become a superlative competitor in the wild. Such worries cannot be dismissed out of hand. But many biologists, impressed by the high degree of mutual adaptation of parts that one finds in evolved and established species, might rather expect nitrogen-fixing wheat to have the same kind of fitness as a man walking on stilts – enhanced in one particular way but crucially handicapped in other ways. This prediction should not be difficult to test empirically should the situation ever arise.

Risks due to unforeseen effects of DNA insertion

Given the essentially random integration of transforming DNA in plants, one must expect some transformants to have suffered alteration of resident genes through insertion of foreign DNA into either coding or control regions. Most such insertions would knock out or severely impair gene function, but in rare cases the effect might be to enhance gene expression, or perhaps cause the gene to be active in parts of the plant where it was normally silent. It would be unfortunate, for example, if engineered potatoes were to produce in their underground tubers the toxic alkaloids that normal potatoes form only in green tissues. In fact, effects of this kind can occur as a result of spontaneous mutation and potato breeders already have to watch out for them.

In so far as random insertion of DNA is a hazard it is one that is already present in at least some plants. Mobile DNA elements, first discovered in maize and probably best regarded as molecular parasites, are indigenous in organisms of many kinds. Their essentially random transposition from one chromosomal site to another is an important cause of the spontaneous mutation that is going on all the time in cultivated as well as wild plant populations. But it seems impossible to cite an example of a cultivated variety being converted into a weed as the result of a single mutation.

Risks from outcrossing to other plants

Gene transfer to weed species

Several crop plants in the UK, and a large number world-wide, are able to hybridize with related wild species that may be growing in the same area. Pollination of a wild species by a genetically engineered crop could result in transfer of an artificially manipulated gene into the wild species. Hence the same range of concerns about possible increased weediness arise again at one remove. Indeed, they gain greater plausibility in that wild species are usually better able to compete than cultivated ones. On the other hand, hybrids between crops and wild species will occur initially only as sporadic single plants, and would only be likely to establish themselves in large numbers if the foreign gene conferred a very marked competitive advantage. We have already considered arguments for thinking that such advantage seems quite improbable as an outcome of most of the deliberate genetic changes envisaged at present, and even less probable as a consequence of unintended random effects of DNA insertion. However, any engineered change in a crop that *was* conducive to invasiveness would be likely to be more so if transferred to a wild species.

The projected crop modification most likely to spread into related weedy species is herbicide resistance. Transfer of a herbicide-resistance gene to wild relatives could cancel the advantage of using the herbicide to which the crop had been made resistant. This would have the same impact as natural selection and spread of a spontaneous herbicide-resistance mutation in the weed population; to judge from precedent, this is quite likely to happen in any case if a particular chemical agent is used persistently.

How many currently important crops are, in fact, sufficiently closely related to wild weedy plants to form fertile hybrids with them? There are many examples world-wide, but the list for the UK is quite short and does not include any of our important cereals or leguminous crops [8]. The cultivated oat, *Avena sativa*, is considered to be a different species from the wild oat *A. fatua*, and the two, though rather alike in appearance, seem to hybridize rarely if at all. The various cultivated beets – beetroot, sugar-beet, spinach-beet, chard, mangold – will all hybridize freely with each other as well as with the wild beet of the seashore when given the opportunity. Wild beet, and the products of its hybridization with sugar beet, can be a serious pest for growers of that crop. The cultivated carrot is placed in the same species, *Daucus carota*, as the wild carrot, and the two can hybridize freely. The wild carrot became a troublesome weed in the eastern USA after its import-ation from England, but does not seem to give much trouble on this side of the Atlantic. The common vegetables cabbage, cauliflower, broccoli, brussels sprouts and kale are all forms of *Brassica oleracea*, but the wild species occurs only locally in the UK, mostly on maritime cliffs. Another species of *Brassica*, *B. napus* or rape, now very extensively grown as a source of oil, is placed in the same species as the root crop swede, but is not closely related to any wild British species that might be regarded as a potential weed. It is, however, a potential weed in itself, a consideration that does not seem to have inhibited its large-scale cultivation.

Gene transfer to other crops
Novel genes could be transmitted from a genetically engineered crop variety to other crop species or varieties for which they had not been intended. In the UK and other temperate countries the obvious possibilities for such transfer are within the beets and the brassicas. It is obviously possible for transfer to occur between different varieties of the same crop, especially where, as in the brassicas, cross-pollination occurs in preference to self-pollination. To the extent that this happens, it will clearly be a nuisance to the plant breeder, but it is not a special hazard of genetic engineering. It is undesirable to have cross-contamination of expensively bred varieties no matter what the sources of their genes. So far as hazard to the environment is concerned, a gene that is judged safe in one crop should usually be safe in another. But there could be special reasons to the contrary in particular cases. To take a hypothetical example, a gene that could be certified as safe for beetroot because it was

expressed in the leaf and not in the root might still be suspect if it looked likely to be transferable to spinach beet.

The analogy with invasion by alien species

An analogy is sometimes drawn between the introduction of a new crop variety obtained by genetic engineering and invasion by a species from another part of the world. Alien plant introductions have certainly had unfortunate consequences on occasion. A plant that is vigorous but tolerable in one country may become a pernicious weed in another – this is certainly the way in which the common English blackberry is regarded in Australia, to take just one example. A high proportion of North American arable weeds came from Europe, no doubt carried over by the early colonists with their crop seed. Weedy invasions seem in general to be from countries with a long history of agriculture into newly opened-up territory, rather than in the reverse direction. Agriculture in the UK has suffered from them only to a minor extent. The list of examples consists mostly of alien species of the family Cruciferae (brassicas, etc.) and only one of these, *Cardaria draba* or hoary cress, is identified in the standard British Flora [8] as a particularly troublesome weed. A more pessimistic review of the effects of plant invasion into the UK has been published [9], but it is questionable whether most of the 'weeds' counted in that survey are seriously troublesome.

Apart from unintentional invasions, very many alien species have been deliberately introduced into British gardens over the last several centuries. Rather few have become naturalized in the wild and very few have become pests. Two exceptions are *Rhododendron ponticum*, which has become unduly dominant in some woodland areas, and the pernicious garden weed *Aegopodium podagraria* (ground elder) which is said to have been introduced as a pot herb by the Romans.

Novel garden hybrids have an excellent safety record. Horticulturalists have made almost every possible hybrid between ornamental wild plants, sometimes taking species that, because of geographical separation, would never have hybridized without human intervention. Some very strikingly novel plants have emerged from this unregulated experimentation, but little or no agricultural or environmental damage.

The analogy between alien importations and genetic engineering can be criticized on the grounds that a tried and tested wild species is likely to have more competitive fitness than a novel derivative of a cultivated plant. It seems likely that the products of genetic engineering will, in general, be less likely than introduced natural species to become weeds. But consideration of the history of plant introductions may lead to some tentative conclusions that may be useful. These are that plants that are vigorous colonizers of open ground in their home territory may well find even more scope for their weediness in another country, but that plants of more specialized habitats (like most of those deliberately introduced for their ornamental value) are very unlikely to become effective weeds through transplantation. Extending this line of thinking to genetically manipulated plants, it seems reasonable to suggest that, while conferring new and possibly advantageous properties on weedy species is to be avoided, genetic engineering of non-weedy cultivated varieties, or transfer of new genes to non-weedy wild species, is very unlikely to produce a new weed. Whether or not a species is weedy, or has weedy potential, is a question best answered by plant taxonomists and ecologists rather than genetic engineers (p. 125).

Regulation of release of engineered plants

In the UK, some of the first genetically engineered organisms to be considered for release from indoor containment have been crop plant varieties, especially potatoes, into which novel genes have been introduced via the *Agrobacterium* Ti plasmid (see p. 49). One case to have been considered, already mentioned above (p. 56), concerned the introduction into potato for expression in the leaves of the pea gene coding for the seed protein lectin. The idea is to improve resistance to aphids.

In deciding the conditions under which it would be safe to grow potato plants expressing the pea lectin gene several questions would seem to be relevant. First of all, would the plants be a hazard in themselves, just growing where planted? The answer is clearly 'no'.

Secondly, would such potatoes be likely to have acquired exceptionally weedy properties by virtue of having acquired a pea protein? The answer is that the cultivated potato is so far from being a competent weed that it is scarcely conceivable that the acquisition of some extra hardiness to insect attack would make it into one.

Thirdly, are there any related wild species to which the engineered potatoes might transmit the pea gene via pollination? The answer is that there are no such species in the UK or Europe.

Fourthly, is there any way the plasmid vector could escape and infect other plants, not necessarily closely related to potato? The answer is that the integrated T-DNA is no more able than any other part of a plant genome to escape and infect – the machinery whereby it got into the potato was left behind in the donor bacterium.

Finally, one might ask whether it would really matter if, through some rare and so far unobserved train of events, the pea lectin gene *were* to be transferred to one or a few plants of another species. The answer to this could only be 'not so far as one can see'.

Presumably it is because of some such combination of reassuring answers that the novel potatoes have indeed been approved for growth in open fields – but only with rather stringent precautions, including the removal of every flower bud. It seems highly probable that, in such apparently straightforward cases as this, precautions will soon be substantially relaxed – as, indeed, they will need to be if genetically engineered crops are to have any commercial future. The same set of questions applied to other cases, however, may not always give such reassuring answers.

References

1 Walden, R. (1988) *Genetic Transformation in Plants*, 138 pp. Open University Press, Milton Keynes.
2 Flavell, R. B. (1989) Plant biotechnology and its application to agriculture. *Phil. Trans. R. Soc. Lond. B* **324**: 526–537.
3 Gasser, C. S. and Fraley, R. T. (1989) Genetically engineering plants for crop improvement. *Science* **244**: 1293–1297.

4 Wilson, T. M. A. (1989) Plant viruses: a tool-box for genetic engineering and crop protection. *BioEssays* **10**: 179–186.
5 Botterman, J. and Leemans, J. (1988) Engineering herbicide resistance in plants. *Trends Genet.* **4**: 219–222.
6 Smith, C. J. S., Watson, C. F., Ray, J., Bird, C. R., Morris, P. C., Schuch, W. and Grierson, D. (1988) Antisense RNA inhibition of polygalacturonase gene expression in transgenic tomatoes. *Nature* **334**: 724–726.
7 Hiatt, A. (1990) Antibodies produced in plants (product review). *Nature* **344**: 469–470.
8 Clapham, A. R., Tutin, T. G. and Warburg, E. F. (1952) *Flora of the British Isles*, 1591 pp. Cambridge University Press, Cambridge.
9 Williamson, M. H. and Brown, K. C. (1986) The analysis and modelling of British invasions. *Phil. Trans. R. Soc. Lond. B* **314**: 505–522.

6 | GENETIC ENGINEERING OF ANIMALS

Manipulation of animal cells

Nearly all the pioneering work on genetic engineering of animals, or at least of mammals, has used mice. Mouse cells have been manipulated outside the body and reintroduced into the female reproductive tract. Two kinds of cells have been used, fertilized eggs and cells taken from very early embryos (embryonic stem cells). The advantage of stem cells is that they can be cultured and multiplied on an artificial growth medium for a certain time without losing their ability for unlimited development if re-implanted into an embryo.

As detailed below, methods have been worked out for introducing DNA into both kinds of cells [1,2]. But, in the case of stem cells, certain kinds of genetic changes can be achieved without direct DNA manipulation. The cells in culture can be exposed to a mutagenic agent and specific kinds of mutants can be selected because of their resistance to certain normally toxic drugs. For example, cells with mutations that result in the loss of the enzyme hypoxanthine phosphoribosyl transferase (HPRT) are resistant to azaguanine because they cannot incorporate this guanine analogue into their nucleic acids. HPRT-deficient cells, forming colonies in azaguanine-containing medium while millions of non-mutant cells remained inhibited, have been implanted into early mouse embryos. On reintroduction into the uterus, these developed in some cases into chimaeric mice whose tissues were a mosaic of normal and mutant sectors, the latter descended from the introduced cells. In some of the chimaeras a proportion of the germ cells were mutant, so the mutation was transmitted to whole animals in the following generation. A further generation of brother-sister mating yielded animals inheriting the mutation from both parents, and these were wholly deficient in HPRT [3].

The point of this exercise was to provide an animal model for cogenital HPRT deficiency in man (Lesch-Nyhan syndrome), which is a particularly unpleasant condition involving compulsive self-mutilation. Somewhat frustratingly, the HPRT-deficient mice turned out to be phenotypically normal. Another kind of mouse disease model, this time obtained through DNA manipulation, is referred to on p. 82.

Introducing DNA into animal cells

Injection into cell nuclei

It was shown about a decade ago that animal cells in culture could assimilate and integrate into their chromosomes DNA supplied in the form of a calcium precipitate. More recently, higher transformation frequencies have been obtained by direct injection by micro-syringe of DNA into cell nuclei. A favoured procedure is to take fertilized eggs and to inject the DNA into one or other of the as-yet unfused male and female germ nuclei (Fig. 6.1a). This is relatively straightforward with mouse eggs but more difficult with eggs of larger animals because their generally greater opacity tends to obscure the target nucleus. These difficulties are, however, being overcome. Only a few per cent of infected eggs survive the treatment, but most of the survivors have the injected DNA integrated into one or more of their chromosomes. Once integrated at the egg stage, the foreign DNA is transmitted to every cell of the developing embryo and is inherited reliably through further generations. Animals that have acquired a new gene in this way are called 'transgenic', and the introduced gene a 'transgene'.

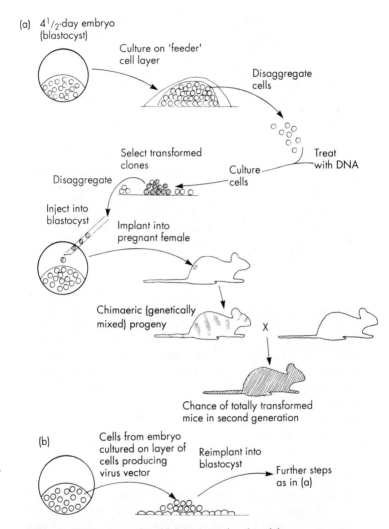

Fig. 6.1 Three ways of making transgenic mice. (**a**) Transformation of cultured embryonic stem cells using a virus vector or vectorless DNA, selection for the desired type of transformants in cell culture, and re-implantation of selected cells into the embryo. (**b**) Infection of whole embryo with retrovirus carrying the desired gene. *Contd. over*

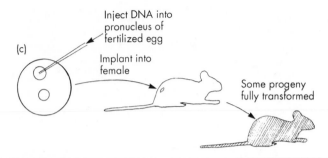

(**c**) Injection of DNA into one of the (as yet unfused) germ nuclei in the fertilized egg. Methods (a) and (b) produce chimaeras – i.e. animals with only some patches of tissue carrying the desired gene; wholly transformed animals can be obtained after two more generations of breeding only if the transformed patches in the chimaera extend to the germ line. Method (c) can give wholly transformed animals in one step. Based on Wilmut *et al.* [2].

An alternative target for DNA injection (Fig. 6.1b) is the nucleus of the embryonic stem cell. Stem cells transformed in this way can be inserted into early embryos to form chimaeras just as outlined above. The disadvantage of this method as opposed to egg injection is that it takes two further generations to obtain a wholly transformed mouse along with a number of untransformed progeny. The advantage is that the stem cells can be manipulated and tested in culture before being committed to the embryo. This is important in connection with gene targeting, considered below.

The use of viruses as vectors

An alternative way of getting foreign DNA into embryonic stem cells is to use a vector derived from a virus genome (Fig. 6.1c). A number of kinds of animal viruses can establish themselves in cells in latent form by integration into host chromosomal DNA. If the virus genome is manipulated so as to include a foreign DNA sequence, that sequence will also be inserted. The principle is just the same as outlined in Chapter 5 for plants (p. 52) and again the best candidates for the role of vector are retroviruses which, though infectious as RNA, make DNA by reverse transcription and insert it into chromosomes of the host cell. The high efficiency of their DNA insertion is the main advantage of retroviruses as vectors. Their limitations are much the same in animals as in plants: the virus genome can accommodate 'passenger' DNA only up to a limited amount and the virus genome has to be disarmed by deletion of some of its vital functions so as to make it incapable of causing disease.

A favoured scheme for disarming virus vectors is to deprive the vector genome of all the genes necessary for reverse transcription and for making infective particles, and then supplying these functions by growing the vector in a special 'helper' cell line. The helper cell contains, integrated somewhere in its chromosomes, an almost full virus genome lacking only a short sequence necessary for its own packaging. Thus in this system there are two complementary crippled virus genomes, the vector and the helper, the first packageable only when helped by the second, which is unable to get packaged itself. The vector particles released in this system contain reverse transcriptase (provided by the helper) and so they are able to make DNA for insertion into the chromosomes of the cells that they infect. But, separated from the helper, the vector should not be able to go through another round of propagation (Fig. 6.2). There is, however, some possibility of recombination between the vector and the helper (or some wild relative) to make a functional virus. Further refinements of vector design could probably greatly reduce the likelihood of such recombination, but it may not be possible to rule it out completely.

Because of these problems, future developments in animal transgenesis are likely to be based mainly on one or other of the available vectorless transformation systems. For genetic engineering of fowls, however, the use of a virus vector seems to be the only method currently available [5].

The problem of gene targeting

In none of the methods for getting foreign DNA into animal cells is there any control over where in the chromosomes the new genes become integrated. This may or may not be important. If a

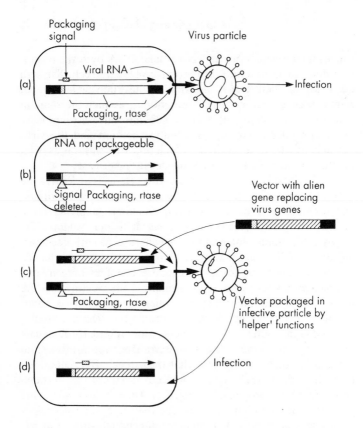

Fig. 6.2 Use of a virus-based vector and a packaging cell line to introduce a transgene into animal cells by infection. (**a**) Cell with integrated retrovirus DNA, transcribed into RNA and packaged in infective virus particles with proteins encoded by the virus DNA. (**b**) 'Helper' cell, with virus DNA deleted with respect to a segment (stippled box) that is essential for packaging of the RNA. Hence the helper produces no infective virus. (**c**) Introduction into helper cells, by artificial transfection, of a constructed vector with virus terminal sequences and packaging signal but with virus genes replaced by the transgene (cross-hatched segment). The transcript of this construct is packaged into infective particles with the virus packaging proteins and reverse transcriptase (rtase) supplied by the helper. (**d**) Cell infected with the vector RNA and transformed by the activity of the transgene after reverse transcription of the RNA into DNA.

transgene is completely alien, with no counterpart in the normal genome, there will be no *a priori* reason for preferring one site of integration to another, provided that the insertion event does not disrupt any essential native gene. But if, as may increasingly happen in the future, it is desired to introduce an improved version of a gene already present, it will probably be an advantage to get the new gene inserted at its normal (homologous) locus, replacing the previously resident version. The new gene will then be more likely to be active at the right time and place and at the right level.

Such homologous recombination events do occur in animal transgenesis, but they are rather rare – perhaps 1% or so of all integrations. If homologous events are required, it is therefore necessary to select the small minority of cells in which they have occurred and to throw away the rest. Here stem cells rather than eggs are the favoured material since they can be grown in culture for a while and tested before any of them are put into embryos. Screening for the relatively rare homologous gene replacements would until recently have been impossibly laborious and time-consuming, but the polymerase chain reaction (referred to already on p. 43) makes it feasible [4]. The principle of the test is illustrated in Fig. 6.3.

What kinds of genes might animal breeders wish to introduce?

Genes for improving the yield of normal animal products

Experience from conventional selective breeding of animals shows that the effectiveness of selection is in general due to the presence in the population of several or many gene differences with cumulative but individually small effects. In the future, by a combination of molecular and statistical analysis, it may be possible to identify and clone some naturally occurring single-gene variants that make particularly significant individual contributions to meat or milk yield, but this is not easy and has not yet been done.

In the absence of information about such genes, animal breeders are a little short of ideas about how to apply genetic engineering to the improvement of livestock. Most attention has

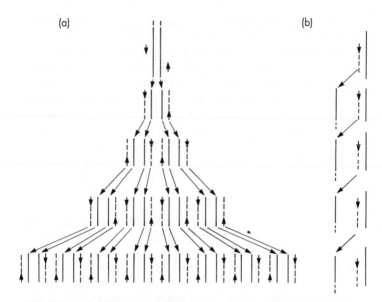

Fig. 6.3   Amplification of DNA by the polymerase chain reaction to test for homologous integration of a gene in a clone of stem cells. A small sample of DNA from the transformed cell clone is used as template for *in vitro* DNA synthesis. Two short pieces of single-strand DNA (short arrows), made to match sequences respectively within the gene (downward-pointing arrow) and within its normal flanking region (upward-pointing arrow), are used as primers. If, as in (**a**), the gene is integrated at its normal site, each newly synthesized strand can itself be used as a template by reverse priming, and the DNA sequence between the priming sites is amplified geometrically to the point of being readily detectable as an abundant DNA fragment of predicted length. If as in (**b**), the transgene is integrated at some non-homologous locus, only one-way rather than reciprocal priming of synthesis of the transgene will occur and amplification will be insufficient for detection.

been focussed on growth hormone, a protein produced by the anterior pituitary gland that is essential for normal growth. In what was probably the first exercise in animal genetic engineering to achieve wide publicity, an additional copy of the gene encoding the hormone was introduced into a mouse under the control of a promoter that caused it to be expressed in the liver at a level that was both high and released from normal control. As a consequence the mouse grew much larger than its litter-mates. The same procedure was tried out on pigs. The result in this case was not larger size but something even more desirable – an increase in the proportion of lean tissue to fat. Unfortunately, a number of other less desirable effects (gastric ulcers, arthritis, dermatitis and renal disease) made it difficult to rear and breed from the modified animals [5].

Interest in increasing the level of growth hormone by introducing extra copies of its gene is limited by the fact that it is much simpler to boost hormone levels by injection. The hormone may itself be a product of genetic engineering of bacteria (see Chapter 2). It is well established that injection of bovine growth hormone (BGH) will increase yields of meat and, particularly, of milk. This use of BGH has been quite extensive in the USA but has been banned in Europe because of concern about possible risks to the consumer from traces of residual BGH in the milk or meat.

Among other proteins that have significant individual effects on animal productivity are the keratins, the major proteins of wool. Keratins are very rich in the sulphur-containing amino acid cysteine, which sheep do not always get enough of in their diet. A project started in an Australian laboratory is to endow sheep with additional genes of bacterial origin that will enable them to synthesize cysteine for themselves from more abundant herbage constituents. Whether this will work remains to be seen. It seems a little unlikely that one could introduce an entirely new metabolic pathway into a complex animal without side-effects, and some of these could be deleterious.

Genes for disease resistance

There has been some interest in the possibility of making livestock immune to virus infections by endowing them with genes encoding virus components. We saw a similar idea with respect to plants in Chapter 5 (p. 59). One possibility is to introduce a virus gene encoding the protein on the surface of the virus particle that attaches to special receptor sites on the surface of the animal cell prior to infection. The idea is that if the animal itself produced an excess of the attachment protein, it could saturate all the receptor sites and exclude the virus.

A more direct strategy would be to endow animals with ready-made genes encoding antibodies specific against the more important disease viruses and/or bacteria. The animal immune system is an exception to the normal rule of constancy of the DNA during development. Through a series of DNA rearrangements, some controlled and some essentially random, the cells destined to produce antibodies generate an immense variety of antibody-encoding genes. The immune response depends upon the selective multiplication of cells, initially a tiny minority, that have their DNA rearranged in such a way as to encode the antibodies needed to counter the particular infection. It is possible in principle to clone rearranged antibody-encoding genes from antibody-producing cells and to use them to make transgenic animals with constitutive immunity against particular pathogens.

Whether these techniques will work remains to be seen. If they do, they could be of very great importance for the health of livestock.

Genes for converting animals to new uses

Production of pharmaceuticals
If the possibilities of using genetic engineering to make animals perform better in their conventional roles seem at present rather limited, the prospect of adapting them to radically new roles seems much more immediate. There is no reason in principle why a gene encoding a valuable and previously scarce pharmaceutical product should not be introduced into an animal together with a promoter sequence that will cause it to be transcribed strongly in some tissue from which it can be readily recovered.

One such project of this type is designed to make sheep secrete human blood clotting factor IX [2]. This is a protein with which sufferers from haemophilia have to be supplied to avoid the risk of severe bleeding. It is at present prepared from human serum, a somewhat risky source because of the possibility of infection with human immunodeficiency virus (HIV or AIDS). Growth factor IX could not be made in engineered bacteria because, in addition to the primary amino acid sequence encoded in the factor IX

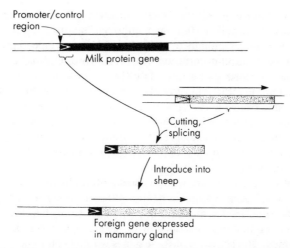

Fig. 6.4 The principle of obtaining expression of a foreign gene in the mammary gland and the production of its protein product in the milk. The coding sequence of the foreign gene (stippled bar) is detached from its own promoter/control sequence and fused to that of a milk protein gene. After introduction into a female animal (see Fig. 6.1), the foreign gene is expressed in the mammary gland and its protein product secreted with the milk. After Wilmut *et al.* [2].

gene, its activity depends on further chemical modifications that occur only in animal cells.

The human gene encoding factor IX was cloned and coupled to a promoter region taken from the sheep gene for a major milk protein. This controlling sequence was expected to switch on the factor IX gene in the mammary gland and cause the factor to be secreted into the milk at a level corresponding to that of the milk protein in a normal animal (Fig. 6.4). If that really worked there would be the prospect of satisfying the world market for factor IX (which needs to be administered only in small doses) from the milk of a relatively small flock of ewes. One may wonder about the effect on the sheep or producing such large amounts of such a highly potent pharmaceutical, but the expectation is that the factor IX would be effectively excluded from other body fluids, since milk proteins are not normally detectable in the bloodstream.

In fact, the expectation of lack of ill effect on the sheep may not yet have been fully tested since, although the project worked in that detectable amounts of active factor IX were indeed present in the milk, the concentration was not nearly as high as had been hoped. The reasons for this relatively low yield are not clear. A considerable effort will certainly be made to tune up the system. The same principle can obviously be extended to the production of any protein factor or hormone for which a cloned gene is available. A second example is the synthesis in the milk of mice of human α-antitrypsin, a product of high promise for the treatment of emphysema.

Substances that are end-products of multi-step metabolic 'pathways', such as vitamins or antibiotics, are much less immediate possibilities for production in animals, since their synthesis would require the introduction of several genes for different enzymes. The activities of these genes would need to be coordinated and the genes themselves linked together so that they would not be split up during hereditary transmission. These requirements could, however, probably be met if commercial motivation were sufficiently strong.

Animals as models for human disease
We saw above how a strain of mice simulating Lesch-Nyhan syndrome was created by random mutagenesis and the manipulation of cells and embryos. It is certain that more animal (usually mouse) models for the study of human diseases of various kinds can be created by DNA manipulation. The first examples are the so-called cancer mice developed at Harvard University. It has been known for a number of years that most, perhaps all, malignant tumours in both man and mouse arise through mutations in genes that, in their normal forms, play essential roles in controlled cell division and growth. Tumour initiation appears to depend on the accumulation of several such mutations in different genes, but if a mouse is created in which one of the mutant genes (oncogenes) is present from birth the chance of the conditions for tumour formation arising during the lifetime of the mouse is greatly increased. The Harvard cancer mice, in fact, harbour two different oncogenes in different strains. Each strain develops cancers at a certain age in a predictable way. When the

two oncogenes are combined in one strain by cross-breeding, the onset of cancer is strongly accelerated.

It is expected that these mice will be in considerable demand for cancer research. They were the subject of the first animal patent application to be granted in the USA; a similar application has had more difficulty in Europe [6].

The extent of the possible benefit

It is difficult to see, at least in the next few decades, any great benefit from genetic engineering to the breeding of farm animals in their conventional roles. In Western Europe and North America the output of meat and milk is in any case already far more than adequate. In the Third World there is great scope for improvement in animal production. Anything that improved disease resistance would be of great benefit, and here genetic engineering may have an important part to play. Its importance, however, may lie more in providing new vaccines (see Chapter 7) than in making resistant transgenic animals.

The possible use of animals as 'living factories' for the production of valuable pharmaceutical products has to be taken very seriously. The market is so great and the competition so intense that every possibility for making new products, or making old ones more efficiently, is likely to be explored. It will be surprising if there is not a significant contribution from genetically engineered animals within the next decade or so. Genetic engineering of microbial cells is not always an alternative. Some therapeutic proteins, such as factor IX, must be produced in an animal system since only animal cells have the metabolic apparatus for making the necessary secondary modifications to the polypeptide chain. If factor IX, for example, really can be obtained in reasonably high concentration in milk, it can certainly be purified from that source more easily than from human serum where it is present only at low concentration and, moreover, with some risk of virus contamination.

The value to medical research of animal models for human disease is fairly obvious. It is true that no animal can be a completely reliable model for a human being, but it is nevertheless true that most of the cellular machinery and controls of development are basically the same in all mammals. What happens in an animal with a particular defect in metabolism or in the regulation of cell division must at least be relevant to investigations of the corresponding human condition.

Objections to genetic engineering of animals

Ecological considerations

Those who question the acceptability of genetic engineering are usually concerned about both animals and plants, but in fact the grounds for concern are rather different in the two cases. Hardly any of the hypothetical risks that we considered in relation to plants can reasonably be held to apply to farm animals. There is obviously no chance of a genetically manipulated race of pigs, sheep or cows running wild and out of control, and there are no wild species with which they could interbreed. Farmed fish such as salmon and trout, were they to be genetically engineered, would be a different case, since they could both escape and interbreed with their wild relatives. Concern has already been expressed about escaping farmed salmon; if sufficiently numerous, they could have a devastating effect on other species of fish, and they provoke anxiety among fly-fishermen about possible mongrelization of the salmon stock. Whether the risk would be any greater if farmed salmon were genetically engineered would, of course, depend on the nature of the genetic modification in each case. Any estimate of added risk greater than zero would probably bring a requirement for containment that would effectively price the engineered fish out of the market. Perhaps that is why so little has been heard about genetic engineering of salmon and trout.

Risks from the use of viruses

In so far as vectors derived from retroviruses are used to introduce genes, it will be necessary to be as certain as possible that they cannot regenerate infectious virus either by themselves or, more likely by recombination with complementary virus sequence present in a 'packaging' cell. It can be argued that, with good vector design, the risk can be made negligible. The chance of any complete virus emerging from the helper cell should be small, and the possibility that it would be more threatening than the already existing virus from which the vector was constructed seems remote. Such judgments depend, however, on confidence in the adequacy of the present understanding of virus structure and function and the competence of the technologists involved. Such confidence will probably be justified, but regulatory agencies will need to be convinced. For most purposes, though probably not yet for transgenesis in poultry, virus-based vectors can in any case be replaced by other methods of gene transfer that will be subject to fewer questions.

Animal rights

Even if the present risk to man from the genetic engineering of animals seems remote, even negligible, we still need to consider whether any of the procedures in use or under consideration is unacceptable from the point of view of animal welfare.

Animal rights is a complex issue about which there is very little clarity. Virtually everyone would agree that animals should not be subjected to injurious procedures without good reason, but how good does the reason have to be? Nobody, in deciding whether or not to pursue a particular objective, knows how to weigh animal against human welfare. Our animal welfare legislation lays it down that animals should be treated as humanely as possible given the uses to which they are being put. But, apart from such minority pastimes as badger-baiting, it is difficult to think of forms of animal exploitation that are forbidden altogether because they are considered to be inherently inhumane. In practice, animals have such rights as we can grant them without too much inconvenience to ourselves.

Let us nevertheless proceed on the basis that the well-being of animals should be given some weight, and ask whether there are any aspects of genetic engineering that might be considered unacceptable on these grounds. The question may be considered under two headings: (i) the means through which transgenesis is achieved in the first place, and (ii) the life of the transgenic animal.

The kinds of human intervention involved in transgenesis – removal and replacement of eggs and embryos and sometimes *in vitro* fertilization – do, from one point of view, constitute flagrant interference with the natural life of animals and arguably violations of animal rights. But it is rather late to raise arguments of this kind and little chance that they could prevail. Artificial insemination, for example, has practically replaced natural coition in modern cattle breeding, undoubtedly with severe detriment to the quality of life of cattle, but this seems to have evoked little concern on the part of the animal-loving public. Nor does anyone except vegans take exception to the traditional production of milk and eggs, both dependent on subversion of normal maternal function.

So far as transgenic animals themselves are concerned, there must be some concern about the welfare of animals engineered to serve as living factories for the production of pharmaceuticals. In theory there need be no adverse effect on the animals if the product is sequestered in the milk, out of contact with other body fluids. But if the tissue control was not absolutely tight and a potent pharmaceutical was made in significant amounts in the liver, for example, then the animal might well be in distress. Some engineered diseases of this kind might occur by accident. We have already noted the multiple handicaps induced in pigs by excessive production of growth hormone. In this case, there was an obvious commercial interest in not persisting with that kind of pig, since it did not thrive or breed efficiently. But even had that not been the case, the condition of the pigs would have seemed unacceptable to many people.

More serious objection might be taken to the creation of mouse or other animal models for human disease. Such animals, almost by definition, are born to a life of handicap and perhaps distress. The ethical question here is much the same as in the use of animals in medical research in general. Opponents of 'vivisection' tend to say that the information obtained from animal experimentation is either useless (because of the inaccuracy of animals as models for humans) or obtainable in other ways such as cell culture. If this were always true it would greatly simplify the issue, but it clearly is not. The experimenters, for their part, would no doubt like to think that their experiments entail no animal distress. But, unless they deny that animals are capable of distress, the most they can honestly say is that they try to reduce it to a minimum. There is no easy escape from the dilemma.

Probably the most potent source of concern about animal genetic engineering is the feeling of many people that the inherent natures of animals – or at least those animals that we feel to be in any way akin to ourselves – ought to be respected and not tampered with. This feeling is sometimes based upon the religious belief that animals are made to God's design, but perhaps more often to the fear that what is done with animals today may be extended to humans at some time in the future. Hopes and fears about the extension of DNA technology to humans are considered in Chapter 8.

References

1 Jaenisch, R. (1988) Transgenic animals. *Science* **240**: 1468–1474.
2 Wilmut, I, Clark, J. and Simons, P. (1988) A revolution in animal breeding. *New Sci.* (7th July), 56–59.
3 Kuehn, M. R., Bradley, A., Robertson, E. J. and Evans, M. J. (1987) A potential animal model for Lesch–Nyhan syndrome through introduction of HPRT mutations into mice. *Nature* **326**: 295–298.
4 Capecchi, M. R. (1989) Altering the genome by homologous recombination. *Science* **244**: 1288–1292; see also *Trends Genet.* **5**: 70–76.
5 Pursel, V. G., Pinkert, C. A., Miller, K. F., Bolt, D. J., Campbell, R. G., Palmiter, R. D., Brinster, R. L. and Hammer, R. E. (1989). Genetic engineering of livestock. *Science* **244**: 1281–1287.
6 No patent for Harvard's mouse? (1989) *Science* **243**: 1003; see also *Science* **245**: 25.

7 VACCINES OBTAINED BY GENETIC ENGINEERING

The principle of vaccination

Our ability to survive infectious disease, whether caused by bacteria, viruses or protozoa such as the malaria parasite, is largely due to our immune system, which responds to infection by producing specific proteins, antibodies and surface proteins of white blood cells, that tend to neutralize the infection. The immune response is especially crucial to recovery from diseases caused by viruses, which, with few exceptions, do not respond to antibiotics of the sort that kill bacteria. Drugs that would kill the virus are likely also to kill the patient. So, generally speaking, one just has to wait for the body's immune system to neutralize the viral infection.

However, it is possible to combat infection by mobilizing the immune system in advance through administration of a vaccine, which may be either an attenuated and relatively harmless relative of the disease agent or a preparation of the protein (antigen) that provokes the immune response. In either case the intended effect is to stimulate formation of antibodies without inducing symptoms of the disease. The former method – the use of a live virus or bacterium – is the more likely to provoke an effective immune reaction but also gives rise to the greater concern regarding safety.

Recombinant DNA technology is becoming very important in the development of new vaccines, both live and non-live.

Live vaccines – past successes

The principle of using a relatively harmless virus to stimulate immunity against a related virulent virus has been long known. The unblemished complexions of milkmaids, so frequently celebrated in English folksong, is now attributed to immunity to smallpox acquired as the result of exposure to cowpox virus. The vaccinia virus, which in our own day has been used systematically throughout the world as a live vaccine against smallpox infection, is thought to have arisen by natural recombination between the cowpox and smallpox viruses. It is itself usually harmless, although there have been adverse, even lethal, reactions to it in a small minority of vaccinated children. It always provokes a strong immune reaction which prevents any subsequent establishment of smallpox in the vaccinated individual; as a result, smallpox now appears to be extinct.

A second major success story is the anti-poliomyelitis vaccine. The live virus used in this case is an attenuated (i.e. weakened and non-pathogenic) strain of poliovirus, thought to differ from the virulent virus through perhaps as many as ten successive mutational changes. This vaccine, too, has been very successful in inducing immunity to its dangerous relative, although it, too, has had rare adverse effects.

Genetically engineered viruses for use as live vaccines

The safety record of the live vaccines used in preventive medicine so far has on the whole been good, and it is generally accepted that the benefits from their use have far outweighed their occasional damaging effects. Nevertheless, live vaccines closely related to highly pathogenic viruses, and perhaps derived from them through only a few mutational steps, are inherently risky to some degree. There is always the possibility, however small, of further mutations restoring virulence. And an attenuated virus may itself produce symptoms in particularly susceptible people. In the early days of the live polio vaccine there were quite serious doubts regarding its safety, but the urgency of dealing with the disease at the time outweighed these misgivings. The promise of genetic engineering in this area is to produce live vaccines that are inherently safe.

Vaccinia-based vaccines

Most of the work that has been done on the construction of new viruses for use as live vaccines has used vaccinia virus as starting material. This virus has the great merit of having already been used successfully on an enormous scale as a vaccine against smallpox, with only rare seriously adverse effects. Through genetic manipulation it is possible to insert into the genome of this rather safe virus a gene from the pathogenic virus coding for an antigenic surface protein. Vaccination with the engineered virus should then induce the formation of antibodies conferring immunity to the pathogen. The vaccinia virus genome is relatively large and contains several sites at which foreign genes can be inserted and expressed without destroying its ability to propagate.

One effective strategy is to splice the foreign gene into a section of the vaccinia genome, cloned in a plasmid, as a replacement for the vaccinia gene that encodes the enzyme thymidine kinase (TK). This modified segment can then be introduced into the complete virus genome by introducing the plasmid into vaccinia-infected cultured cells and selecting for recombinant viruses lacking TK [1]. Such recombinants can be selected by virtue of their ability to propagate in the presence of the drug bromo-deoxyuridine, which depends upon TK for its toxic effect. In the recombinant vaccinia genome the foreign gene is transcribed under the influence of the TK promoter (Fig. 7.1).

A single inserted gene encoding an antigen of a pathogenic virus should seldom or never convert vaccinia into a serious pathogen, but this obviously has to be checked carefully in every case. The loss of the TK gene reduces the virulence of vaccinia virus [2], so antigen-encoding TK-negative recombinants may actually be safer than the vaccinia used against smallpox.

Progress has already been made towards live vaccinia-based vaccines against hepatitis B virus (HBV). The gene encoding the main HBV surface antigen was inserted into vaccinia and rabbits infected with the recombinant virus were shown to produce antibodies against HBV [3]. A similar strategy has been used to immunize rabbits against Epstein–Barr virus, the causative agent of infectious mononucleosis in humans, and the malaria parasite [4]. Research is also proceeding on a live vaccine for combating foot-and-mouth disease in sheep.

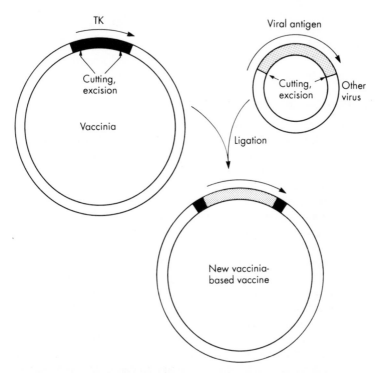

Fig. 7.1   Principle of construction of a recombinant vaccinia virus for use as a vaccine against another virus. The coding sequence for an antigenic component (e.g. coat protein) of the second virus (stippled bar) is substituted for that of a vaccinia gene (e.g. the gene for thymidine kinase, TK, black bar) in such a way that it can be efficiently transcribed from the vaccinia promoter. The antigen gene does not confer pathogenicity on the vaccinia virus, and the loss of TK function handicaps the recombinant vaccinia virus to a moderate extent.

Other viruses for vaccine construction

Vaccinia is not the only virus being considered as starting material for vaccine production. For example, another virus of the pox group, fowlpox, has been used in pilot studies on a possible anti-rabies vaccine. A gene coding for a surface component of rabies virus was inserted into the fowlpox genome. When used to vaccinate a range of experimental animals, the hybrid virus induced the formation of anti-rabies antibodies in all of them. Fowlpox only causes disease in birds, and the experimental animals did not show disease symptoms as a result of the vaccination [5].

Infectious spread of live vaccines

When live viruses are used in vaccination programmes, it is difficult to be sure that they are never accidentally transmitted to people who had not intended to be vaccinated. It is generally believed that some unintentional transmission of vaccinia and attenuated poliovirus did indeed happen during the campaigns against smallpox and poliomyelitis. It is even likely that such transmission, leading to more widespread immunity in the population, contributed substantially to the success of the anti-polio campaign.

Involuntary immunization by infection of the population at large, though perhaps accepted in the case of polio, is likely always to be a controversial objective for a human vaccination programme. It is, however, the overt aim of projects for controlling rabies in wild animals. The idea is to leave bait for the animals containing live vaccinia or cowpox virus engineered in either case to produce a rabies virus antigen. It is hoped that enough animals will take the bait to start a vaccine epidemic that will spread through the entire animal population and eliminate, or at least usefully reduce, the endemic rabies virus. A first field trial of such a project was approved by the state of Virginia in 1989 (having previously been turned down by South Carolina) [6]. The test area is an island but presumably, if the test is successful, the next step will be to try to establish the vaccine over a broader area. The first field trial of a similar scheme to control rabies in foxes on the continent of Europe was made in Belgium in 1988.

Live vaccines for protection against bacterial infections

Intestinal infections, caused by such bacteria as *Salmonella typhi* (typhoid fever), *Shigella dysenteriae* (dysentery), *Vibrio cholerae* (cholera) and pathogenic strains of *Escherichia coli* (diarrhoea), still pose huge problems in Third World countries. They can in principle be controlled by oral vaccines consisting of attenuated bacteria. 'Naturally' attenuated bacterial strains have been extensively used for a number of years. Non-invasive live *Salmonella* vaccines that can give lasting immunity against typhoid already exist. Genetic engineering extends the possibilities [7]. It makes it possible to attenuate a pathogenic bacterium by irreversible deletion of defined genes. Candidates for deletion would include genes coding for essential components of the toxin that might account for the pathogenicity, and genes essential for metabolic processes needed for long-term viability of the bacterium but not for its immediate triggering of the immune response.

Another possible approach is to insert an immunity-provoking gene from the pathogen into a related but non-pathogenic species that may itself be handicapped in various controlled ways. These developments are still at an early stage, but their potential for human health is clearly very great.

Risks of live recombinant viruses

A proposal to add a gene from a dangerous pathogen to a normally benign but potentially infectious virus or bacterium must raise questions about safety. In order to assess the prima facie risk of such a procedure, it is necessary to understand the role that the added gene plays in the life history of the pathogen. The ideal gene would be one that encoded a packaging protein that played no part in the propagation or targeting of the virus after it has entered the host organism. Unfortunately it is not safe to assume that a surface protein of the virus particle is just inert packaging – haemagglutinin of influenza virus is an example to the contrary. An obvious kind of gene to be avoided would be one responsible for targeting the virus to the particular tissue in which the pathogen does its damage – hepatitis virus to the liver, for example. There would be a real possibility that such a gene could make a normally harmless virus dangerous. Genes coding for bacterial toxins should clearly not be used in vaccine construction unless, as in the case of cholera, more than one component is required for fabrication of the toxin, in which case a gene for one component alone may pose no threat.

A possibly safer alternative to using a complete gene from a pathogen to generate an immune response is to identify the particular region of the gene (coding for just a part of the protein product) that has the greatest immunogenic effect. One can then isolate (or indeed synthesize) that part of the gene and insert it into the DNA sequence of the vaccine at a site at which it can be efficiently expressed. This approach has been used, for example, in experiments directed towards the development of a vaccine for foot-and-mouth disease of sheep. It seems quite unlikely that a minimal protein fragment selected purely on the basis of its ready recognition by the immune system would present any risk by itself even if the whole protein to which it normally belonged had some role in pathogenicity.

Safe vaccine design requires an adequate understanding of what it is that makes the pathogen pathogenic. The designs at present developed may well already be safer than some successful vaccines were when first tested, but there is certainly room for improvement in the understanding of pathogenicity. And whatever theoretical arguments there may be for the safety of new live vaccines, they will always need empirical confirmation. This will have to be by tests on experimental animals in the first instance and ultimately (since animal models are not absolutely reliable) on consenting human patients.

Non-living vaccines

Many of the vaccines in current use consist of preparations from dead pathogenic bacteria. Examples are the preparations used to induce resistance to tetanus and bacterial pneumonia and meningitis. There is less chance of an accident if the vaccine consists not of whole killed bacteria, but of a single antigenic constituent of the bacterial cell. For example, a modified form of tetanus toxin (toxoid) is used as the vaccine against tetanus.

The same principle can be used for immunization against

viruses. Rather than use the whole killed virus, it may be more effective and (because viruses tend to survive any but the most extreme treatments) also safer to use as vaccine only an immunity-inducing protein component of the virus without the viral nucleic acid. Such a viral subunit vaccine can be most safely and efficiently obtained by cloning and obtaining expression of the appropriate virus gene in a harmless bacterium such as *E. coli* K12.

One example of this approach is the development of a vaccine against hepatitis B virus (HBV). For several years now it has been known that an effective vaccine can be obtained by purifying the major HBV protein, that encapsulates the DNA in the infective virus particle, from the serum of infected individuals. The same protein can now be obtained more efficiently from a bacterial strain engineered so as to express the appropriate HBV gene, without any need for a supply of sufferers from the disease [8]. The vaccine is inherently safe – without the HBV DNA there is no way that it can regenerate the virus. The process of making it is also free of obvious danger, and would in any case be contained within an industrial plant governed by the safety regulations outlined in Chapter 3.

The same principle has been used to prepare a prototype vaccine against human immunodeficiency virus (HIV, AIDS virus) [9]. In this case the vaccine consists of a fragment of a virus protein produced by suitably engineered yeast cells, and the first trials on human volunteers to check its effectiveness in provoking an immune response are reportedly about to begin in the UK as this book goes to press.

The immune response to subunit vaccines is generally less vigorous than it is to live vaccines, but it can be boosted by injecting the vaccine together with other substances called adjuvants. Experiments on chimpanzees indicate that effective protection against hepatitis can indeed be achieved in this way.

Conclusions

It is arguable that the production of new vaccines is likely to be the most beneficial application of genetic engineering. Because it necessarily involves operations with dangerous pathogens it also carries the clearest risks. But because the risks are relatively well defined and understood they can be more satisfactorily guarded against than some of the more nebulous dangers discussed elsewhere in this report. There is every prospect that vaccines developed through genetic engineering will be safe as well as effective – safer, probably, than some of the presently acceptable vaccines made by methods little different from those used by Louis Pasteur.

References

1 Mackett, M., Smith, G. L. and Moss, B. (1984) General method for production and selection of infectious vaccinia virus recombinants expressing foreign genes. *J. Virol.* **49**: 857–864.

2 Buller, R. M. L., Smith, G. L., Cremer, K., Notkins, A. L. and Moss, B. (1985) Decreased virulence of recombinant vaccinia virus expression vectors is associated with a thymidine kinase negative phenotype. *Nature* **317**: 813–815.

3 Cheng, K.-C., Smith, G., Moss, B., Zavala, F., Nussenzweig, R. and Nussenzweig, V. (1986) Expression of malaria circumsporozoite protein and hepatitis-B virus surface antigen by infectious vaccinia virus. *Vaccines* **86**: 165–168. Cold Spring Harbor Laboratory.

4 Mackett, M., Arrand, J. R., Reith, R. and Williamson, J. D. (1986) Characterisation of vaccinia virus recombinants expressing the Epstein-Barr virus membrane antigen gp340. *Vaccines* **86**: 293–297. Cold Spring Harbor Laboratory.

5 Taylor, J., Weinberg, R., Languet, B., Desmettre, P. and Paoletti, E. (1988) The use of fowlpox virus vectors to vaccinate non-avian species. *The Release of Genetically Engineered Micro-Organisms* (eds M. Sussman, C. H. Collins, F. A. Skinner and D. E. Stewart-Tull), pp. 89–103. Academic Press, London.

6 Virginia OKs rabies vaccine test. (1959) *Science* **243**: 1134–1136.

7 Bloom, B. R. (1989) Vaccines for the Third World. *Nature* **342**: 115–120.

8 Murray, K. (1988) Application of recombinant DNA techniques in the development of viral vaccines. *Vaccine* **6**: 164–174.

9 First UK trial of AIDS vaccine approved. (1990) *Nature* **346**: 303.

8 | APPLICATIONS TO THE HUMAN GENOME

Probing for variation in the genome

Detection of restriction fragments

Central to all applications of DNA technology to humans is the detection of differences between individual people. All genetic differences are, at root, differences in the base sequence of DNA. Variation in DNA sequence is detectable when it results in changes in the sizes of the fragments into which the DNA is cut by restriction endonucleases (see p. 15). Restriction fragments can be separated according to size in a gel in an electric field (electrophoresis), and blotted from the gel on to a membrane which can then be probed with a radioactively labelled DNA fragment that will hybridize to the sequence that is being sought. The radioactivity then reveals the positions, and hence the sizes, of the restriction fragments in which that sequence is present. The essence of the procedure is summarized in Fig. 2.6 (p. 18).

RFLPs and DNA fingerprinting

Variation between individuals in sizes of restriction fragments can arise in several ways, explained in Fig. 8.1. When one particular variation in size occurs in a substantial proportion of

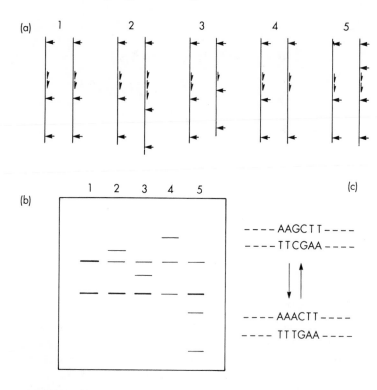

Fig. 8.1 How restriction fragment polymorphisms (RFLPs) can arise. (**a**) Possible variations in a segment of chromosomal DNA with two sites (horizontal arrows) cut with a particular restriction enzyme; between two of the sites there is initially a duplicated segment. The cases illustrated are: (1) ancestral diploid condition; (2) additional repeat in one of the two homologous chromosomes; (3) loss of one repeat in one homologue; (4) loss of a restriction site; (5) gain of a restriction site. (**b**) Effects of these variations on the fragment pattern after restriction digestion and fractionation in an electrophoretic gel (see Fig. 2.6). (**c**) Losses and gains of restriction sites can occur by single base-pair mutations; the example shown here is loss and gain of the 6 base-pair palindromic sequence (AAGCTT) that is cut by restriction enzyme *Hind* III. Changes in number within tandem arrays of repeats occur relatively frequently, probably as a result of natural reciprocal exchange.

the population it is called a restriction fragment length poly-morphism, or RFLP. A particularly rich source of RFLPs in humans is a highly repetitive short sequence that was discovered several years ago and referred to, for reasons that need not concern us, as minisatellite. This sequence is scattered through-out the genome in repetitious head-to-tail arrays. The positions of these arrays are fairly constant, but the number of copies in each array is highly variable between individuals. This means that a minisatellite probe reveals many different RFLPs. The patterns of minisatellite fragment sizes are so diverse that two individuals virtually never share the same pattern.

Because of its discriminatory power, a minisatellite RFLP pattern is referred to as a 'fingerprint'. Unlike an ordinary fingerprint, however, it provides evidence not only of individual identity but also of family relationship. Parents will transmit about half of their DNA fingerprint to each of their children (a different half in each case), and full brothers and sisters will also have about half of their fingerprint in common by virtue of their common parentage. These are far greater degrees of similarity than could occur between unrelated individuals (Fig. 8.2 shows an example) [1].

DNA fingerprinting has already found important uses, both in criminal investigations and in the confirmation of family re-lationships for the satisfaction of immigration officers. In the forensic context, even a small blood stain or a small residue of sperm can yield enough DNA for a fingerprint. Such evidence has already led to a conviction in a rape/murder case in the UK. When the technique is expertly applied to adequate samples it provides an unequivocal identification, but, like any forensic technique, it can be pushed beyond its limits, especially when the DNA sample is extremely small or of poor quality. In the USA, where some optimistically interpreted DNA evidence has recent-ly been successfully challenged in the courts, there has been a growing awareness of the need to enforce high technical standards in the application of the technique [2].

In the immigration context, where there need be no problem about getting adequate samples of DNA, the fingerprinting evidence seems on the whole to have worked to the advantage of immigrants attempting to bring their families into the UK; it has provided evidence of family relationship that the immigration officers might otherwise have disbelieved.

RFLPs can be revealed in many different human DNA sequences. Any randomly selected fragment of a human genome, when used as a probe for its equivalent in somebody else's genome, has a significant chance of revealing a difference. These variations very frequently have no consequences in themselves because they affect regions of the DNA that lack essential function, but, as we see below, they can be of great assistance in locating important genes to which they happen to be linked.

The inheritance of single-gene defects

Many human diseases have some genetic component in their causation. Of these a minority, but still a large number, can each be attributed to a single defective gene [3].

Most defective genes are recessive in that the function that they fail to supply can be adequately met by a single undamaged gene copy inherited from one parent only. Deleterious recessives are present in populations mostly in carrier individuals – heterozygotes – who each have one defective and one normal gene copy and display no adverse effects. Heterozygotes transmit their deleterious gene to one-half of their offspring on average; only if a child has the misfortune to inherit it from both parents will he or she exhibit the disease. Cystic fibrosis, with a carrier frequency of about 1 in 22 and an actual incidence of around 1 in 2000 births, is by some margin the most common recessive congenital disease in the UK, but numerous others occur at lesser frequencies.

Deleterious genes carried on the X-chromosome, which males inherit only from their mothers and thus possess only in single copy, are usually recessive in females but always show their effect in males. One of the important examples of a sex-linked condition is haemophilia – failure of blood clotting – a condition that afflicts about 0.005% of males in the UK.

Defective genes with dominant effects are understandably very uncommon; if severely deleterious throughout life they will usually prevent the individuals carrying them from leaving progeny. But one important example, Huntington's chorea, persists at an important frequency (about 0.01%) because its

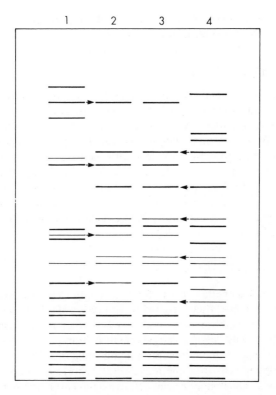

Fig. 8.2   DNA 'fingerprints', showing resemblance between relatives. DNA samples from mother (1), father (4) and identical twin children (2 and 3) were digested with a restriction enzyme, run out on an electrophoresis gel, blotted and probed with 'minisatellite' DNA (for blotting method see Fig. 2.5). Note that the parents differed in many DNA fragment sizes and that the identical pattern of the twins is derived about half from one parent and half from the other (arrows). The children's DNA yielded no fragments that were not also found in one or other of the parental samples. A real example after Jeffreys *et al*. [1].

symptoms (progressive deterioration of the nervous system) do not usually set in before middle age.

Identifying defective genes

Wherever a congenital disease is due to a single defective gene there is an obvious interest in methods that would permit its detection in non-symptomatic carrier individuals and in fetuses. DNA from potential parents is easy to obtain from blood samples. In any pregnancy where both parents are shown to carry the same recessive defect, there is a 1 in 4 chance that any particular child will be affected. All disease due to single gene defects could in principle be avoided if couples at risk either abstained from having children or were able to select in favour of normal births.

Methods of antenatal diagnosis

Given that a method exists for detecting a defective gene by analysis of DNA, there are two possible ways of selection before birth. The established method is selective abortion. Sufficient cells for a DNA sample can be obtained from a fetus, either by withdrawing a sample from the amniotic fluid (amniocentesis) or by sampling cells from the non-essential chorionic villus tissue. Chorionic villus sampling can be carried out earlier in pregnancy than amniocentesis (8–12 weeks as compared to 16 weeks) but at present it carries a greater risk of inducing abortion (2–3% compared with 0.5%) [4].

Embryo selection
This alternative is likely to be increasingly favoured, since it avoids the trauma of abortion. A number of the mother's eggs are fertilized outside the body and allowed to develop in culture to the eight-cell stage of embryonic development. A single cell is removed from each embryo and DNA from it amplified to testable amount by the polymerase chain reaction (see Fig. 6.3 on p. 78). One or two selected embryos can then be re-implanted in the mother, and, if all goes well, the pregnancy then proceeds normally and results in either single births or twins [5,6].

In the eight-cell embryo the cells are still all equivalent in potential for development, and so removal of one cell causes no damage that cannot be repaired by division of the remaining cells. The main current limitation of the method is the relatively low success rate of the *in vitro* fertilization/re-implantation procedure; this is said to be about 10% overall, though it seems to be higher in some of the leading clinics and is no doubt likely to improve further. In a recently reported trial, pregnancy was achieved for 3 out of the 5 couples treated, though not all the successes were obtained at the first attempt [6]. In this first trial, female embryos were selected so as to avoid sex-linked disease (Lesch-Nyhan syndrome, Duchenne muscular dystrophy or other sex-linked disorders), which would have occurred in any male child with 50% probability in the families concerned.

Screening by the polymerase chain reaction for DNA sequences specific for the Y (male) chromosome is relatively straightforward. Selective amplification of sequences diagnostic of defective genes themselves would be more difficult but should become possible, at least for some gene variants, as more is learned about their precise sequences. This would permit extension of embryo selection to autosomal (non-sex-linked) disorders, and, in the case of sex-linkage, discrimination not only between normal and potentially defective male embryos but also between female embryos carrying the defective gene and those free of it.

The use of linked RFLPs as markers for defective genes

By following the transmission of RFLPs through successive generations it is possible to assign them to a number of different linkage groups that can be shown to correspond to different chromosomes. If two RFLPs – let us represent the two pairs of alternative fragment sizes as A/a and B/b – are present in the same individual and belong to different linkage groups, then at the reduction of chromosome number from diploid to haploid (meiosis) prior to germ-cell formation, the two differences will be assorted independently, so that germ cells carrying the combinations AB, Ab, aB and ab will all be equally frequent. But if they are in the same linkage group (i.e. linked on the same chromosome), there will be a majority either of AB and ab or of Ab and

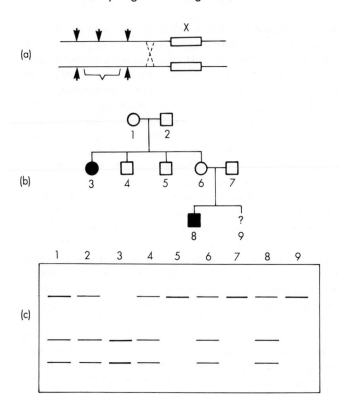

Fig. 8.3 Hypothetical example of the use of a linked RFLP to predict the inheritance of a recessive deleterious gene; the effect of recombination. (**a**) Two homologous chromosomes found in a particular family: the upper carries a deleterious mutant gene (X) linked to a marker segment (RFLP) characterized by three sites cut by a certain restriction enzyme, and the lower has the normal gene linked to an alternative version of the RFLP with only two restriction sites. (**b**) Family pedigree (females conventionally shown as circles and males as squares). Filled symbols indicate that the individual is affected by a disease because of inheritance of a defective gene from both parents. The question is whether the fetus shown as ? is potentially affected by the disease or not. (**c**) Electrophoretic gel, blotted and probed with the DNA segment

aB germ cells, depending on which combinations were inherited from the previous generation.

The fact that the minority combinations (recombinants) occur at all is due to the occurrence of a limited number of more or less randomly placed exchanges (crossovers) between corresponding segments of paired chromosomes during meiosis. The frequency of recombination shown by pairs of linked RFLPs is an index of their distance apart on the chromosome – the higher the recombination the greater the distance. Using data of this kind, RFLPs can not only be assigned to chromosomes but also mapped in a unique linear sequence along each chromosome. It is possible to include in the chromosome map any other simply inherited difference that is being inherited along with the RFLPs.

What clinical geneticists wish to find, and increasingly do find, are RFLPs linked to the genes involved in inherited disease. Then they can use an RFLP, for which they have a probe, to predict the presence or absence in a fetus of a linked defective gene that they could not detect directly. In order to make such predictions one has to determine which of two alternative restriction fragment sizes was linked to the disease gene in at least one of the parents from whom it may or may not have been inherited. In the population as a whole the cumulative effect of crossing-over in every generation is to randomize the combinations of disease

bracketed in (**a**), of DNA samples from members of the family Representing the normal gene as A and defective forms of the gene as a and a', one can predict the following genotypes: 1 Aa, 2 Aa, 3 aa, 4 Aa, 5 AA, 6 Aa, 7 Aa', 8 aa', 9 $A'A$ or a'. Note that the defective gene a', from a different family, happens to be coupled to the two-site version of the RFLP: hence the uncertainty as to whether the fetus (9) has inherited the dominant A or a' from its father. In any event, (9) is predicted to have inherited the dominant A from its mother and thus to be free of the disease. The degree of certainty of the prediction depends on the closeness of the gene to the diagnostic RFLP. A one per cent frequency of crossing over (dashed cross in **a**) will imply a corresponding chance of a false prognosis.

genes and RFLPs. A particular association will only hold up in a particular family and even then only with a certain probability. For example, a 1% frequency of crossing-over between the defective gene and the RFLP would imply a 1% chance of misdiagnosis, which could mean in practice the abortion of a normal fetus or the birth of a defective one (Fig. 8.3).

How close the linkage has to be in order to justify the use of RFLPs as the basis for a decision to terminate a pregnancy is a matter of feeling and judgement. It would obviously be more satisfactory if probes could be obtained to identify defective genes directly.

Direct identification of defective genes

A probe for a particular gene can be obtained fairly easily when the gene in question is abundantly transcribed into messenger RNA in a particular tissue. For example, cDNA probes, reverse-transcribed from messenger RNA, have long been available for the genes encoding the globin components of haemoglobin.

Sickle-cell haemoglobin, detected originally as a tendency for the red blood cells to collapse under conditions of reduced oxygen supply and associated with a severe anaemia, is due to a single amino acid substitution in β-globin, due in turn to a single base-pair substitution in the β-globin gene. Only people homozygous for the mutant gene are anaemic; heterozygous carriers have practically no tendency to anaemia and have considerable resistance to malaria. In countries where malaria has been endemic, and in ethnic groups such as American blacks whose ancestors came from such countries, the sickle-cell mutation is present at relatively high frequency.

It so happens that the most widespread sickle-cell mutation occurred in a β-globin gene that was immediately flanked by an unusual RFLP (Fig. 8.4). The unusual restriction site is so close to the sickle-cell mutation as to be virtually inseparable from it by crossing-over. In this case, therefore, there is virtually no ambiguity about the prognosis based on the RFLP analysis.

It is possible in principle for the same mutation as causes a congenital defect to create or abolish a restriction site in the DNA. There is also the possibility that it is the consequence of a

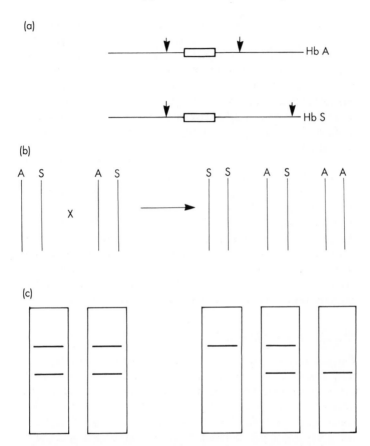

Fig. 8.4   Use of a restriction fragment length difference in a human pedigree to predict sickle-cell anaemia in an embryo. (**a**) The positioning of restriction sites in the usual normal haemoglobin ß-chain gene (HbA) and in the usual sickle-cell variant in the American population. (**b**) The possible genotypes (*SS, AS, AA* with 1:2:1 probability) of children of parents both of whom are of constitution *AS*. (**c**) Patterns of restriction fragments separated by gel electrophoresis, visualized by blotting and radioactive probing. In this case the prognosis is highly reliable since the linkage between the haemoglobin ß-chain gene (Hbß) and the restriction site marker is extremely close (about 1 kb of DNA, corresponding to a crossover frequency of the order of 0.001%). After Weatherall [3].

DNA deletion or insertion large enough to make an appreciable difference to the length of a restriction fragment. In such cases the mutation is directly detectable as an RFLP. Such good luck is unusual. However, with the refinement of the techniques for hybridizing DNA probes to the target genes, it is becoming possible to detect very small differences in the firmness of binding of a probe. Even a single base-pair change may be detectable using a fairly short probe designed to recognize the precise sequence within which the mutation lies. So, given precise knowledge of the nature of the mutation being looked for, it is probably possible to make a probe for it.

The problem is that the information needed in order to make a probe for a gene comes only after the gene or its function have been identified. The transcripts and protein products of many of the genes involved in important congenital disorders have not been identified, so there is no immediate prospect of making probes for them. This has until recently been the situation, for example, with regard to cystic fibrosis, the gene for which has been the object of an international hunt for several years.

'Walking' to genes from RFLP sites

If a gene is closely linked to an RFLP, it may be possible to get to it by a 'walk' along the chromosome using the RFLP as starting point. The principle of chromosome walking is to use two or more different restriction enzymes to generate overlapping restriction fragments. A fragment corresponding to the RFLP that one is using as starting point (call it fragment A) is used as a probe to pick out of a genomic library a restriction fragment B with which it overlaps on one side. Fragment B is in turn used as a probe for the isolation of fragment C, which overlaps with B but not with A.

Proceeding in this stepwise fashion, one can proceed along the chromosome, several tens of kilobases at a time if restriction enzymes that generate very large fragments are used. Some stretches of chromosome may be difficult to walk across, usually because they contain a repetitious sequence that does not provide a specific probe to guide the next step. Such difficult stretches can often be passed by a procedure known as 'jumping'. This depends upon cloning sequences derived just from the ends of extremely

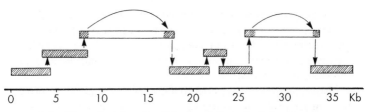

Fig. 8.5   The principle of 'walking' and 'jumping' along a chromosome, starting at a site for which one has a probe, in the hope of reaching a linked gene of interest for which no probe is initially available. The initial probe fragment (left-hand end of diagram) is used to identify and clone an overlapping fragment to the right, which is then used in turn to identify another fragment still further to the right, and so on. Straight arrows indicate use of one fragment to probe for another. The 'jumping' procedure (curved arrows) involves the cloning of very long fragments and the subcloning of only their terminal sequences for use as probes. In this way it is possible to bypass difficult regions, especially those with dispersed repetitive sequences which would be too unspecific to use as probes. Steps in a walk are usually of the order of 1–5 kb, but jumps can be up to 50 kb or more. A notable example of the use of these methods is the recent 500-kb walk/jump along human chromosome 7 from the site of an RFLP to the gene that, in mutant form, causes cystic fibrosis [7].

large fragments (50 kb or more) and passing over the sequences in the middle (Fig. 8.5).

This strategy can work but is usually laborious and fraught with problems. The distances to be covered can be very long. A relatively close linkage (in genetic terms) of 1% recombination corresponds on average to a million or so base-pairs. There is also the problem of knowing whether the walk is proceeding in the right direction from the starting point. A good situation is to have two RFLPs, one on each side of the gene of interest. Then, if the walk has extended from one to the other the direction must have been correct and the gene must have been passed somewhere along the way. Even so, the gene may still not have been recognized, since there may well have been no prior information as to what it should look like.

The greatest success so far of the walking-jumping approach is the cloning of the gene involved in cystic fibrosis [7]. A mainly

Canadian group walked and jumped approximately 500 kb along human chromosome 7 before coming to the gene (which, incidentally, turned out to be of enormous size – 250 kb, consisting mostly of introns). They identified their destination through three pieces of evidence.

Firstly, the region they arrived at had a lot of similarity (shown by mutual probe recognition) to sequences present in other mammalian genomes, suggesting that it served some function of very general importance. Secondly, it was recognized by a DNA probe obtained from a messenger RNA produced in a number of tissues, such as lung and sweat gland, where cystic fibrosis symptoms are manifest. Finally, and most convincingly, most cystic fibrosis patients, but no unaffected persons, were found to be homozygous with respect to a certain small deletion in the coding part of the identified gene. The fact that some cystic fibrosis patients did not have the deletion could be easily explained by the presence in the population of more than one kind of mutation in the gene.

The identification of the cystic fibrosis gene involved an enormous amount of patient work. The provision of a data bank that will make easier future enterprises of this type is a main objective of the human genome project, discussed below (p. 113).

Prospects for gene therapy

So far in this chapter we have been discussing uses of human DNA clones for diagnosis. These uses are already with us. More speculative, and more controversial, is the possibility of human transgenesis for therapeutic purposes.

Genetic repair of somatic cells

Many human congenital diseases are due to defects in genes whose normal functions are to produce particular known enzymes or other proteins in particular tissues. The example of sickle-cell anaemia, due to the production of a defective kind of haemoglobin, has already been referred to (p. 106). Another kind of blood disease, thalassaemia, results from a failure of the bone-marrow cells (reticulocytes) that are destined to develop into red-blood cells to produce adequate amounts of haemoglobin because of a defect in one or other of the major globin genes. One kind of immune deficiency is due to a failure of lymphocytes (white blood cells), that also develop from the bone marrow, to form the enzyme adenosine deaminase, again because of a single gene defect. Phenylketonuria, a relatively common single-gene defect associated with mental retardation, is due to a lack of the enzyme phenylalanine hydroxylase in liver cells. The objective of gene therapy would be to restore the normal gene in each case to the particular tissue in which its activity was needed [8].

At present, bone-marrow cells are the only ones to satisfy two essential conditions for therapeutic gene manipulation – they could be withdrawn from the patient in reasonable numbers and then, after treatment with DNA, they could be put back where they came from by injection. Even so, there are still technical problems.

First, there is as yet no sufficiently efficient way of introducing DNA into these cells other than through the use of an integrating virus-derived vector, and this raises questions about safety of the vector. The safety of virus-based vectors was discussed in Chapter 7. It may well be that careful design, together with extensive experience of their use in animal systems, will be able to give sufficient reassurance on this point.

Secondly, there is the problem of ensuring proper control of expression of the introduced gene. It is difficult to see how it could be targeted to its normal locus without cell culture and selection (see p. 77) which is not feasible with bone-marrow cells. So the gene would have to be equipped with the necessary control sequences before introduction into the vector, and this might make it larger than the vector could accommodate.

Finally, there is the problem of populating the bone marrow with a sufficient number of cured cells in competition with the defective cells already there. The latter could, indeed, be largely eliminated by X-rays, in which case the re-implantation of cured cells would be akin to the procedure already used to rescue victims of serious nuclear reactor accidents. This somewhat drastic solution to the problem might be avoided if the cured cells were made inherently more competitive than the uncured ones, but is not clear how this could be done.

The problem addressed in the last paragraph could be much

less acute in cases where only a small fraction of normal gene activity was needed in order to ameliorate the disease condition. For example, it is thought that as little as 3% of the normal level of adenosine deaminase may be enough to cure the type of immunodeficiency associated with lack of this enzyme. If so, a comparably small proportion of cured white blood cells might suffice. A recent news item (*Science* **249**: 1372) reports that, in the USA, this form of treatment is already being tried on one patient.

What are the prospects of curing cells in other parts of the body such as the liver, where so many essential enzymes, including phenylalanine hydroxylase (deficient in phenylketonuria) are made? One idea is to attach the DNA to a molecule that will carry it to the appropriate cell type. A few years ago there was a report [9] of the use of a protein whose natural function is to insert itself into the cell membranes of liver cells. Chemical linking of a gene to this protein, followed by its injection into the bloodstream, led to the targeted delivery of the DNA and the expression of the activity of the transgene in the liver but not in other organs. Results of this kind give real hope that somatic cell gene therapy may indeed be possible, even if only for a few diseases.

Repair of the germ line

Somatic cell therapy would aim only to ameliorate the disease in the individuals under treatment. The patients, even if themselves cured, could still transmit the defective gene because there would have been no curing of the germ cells. A far more ambitious objective of human genetic engineering would be to put the genetic defect right in the germ cells too and so remove the risk to future generations.

The prospect of realizing this objective in any ethically acceptable way seems at present extremely remote. To get the therapeutic gene into the germ line it would have to be introduced into fertilized eggs or very early embryos, so that it was present in all or most cells of the emerging human being. It would not be known until long after the operation whether the experiment had worked and, if so, whether all the germ cells had been cured or only a proportion of them. Furthermore, unless embryonic stem cells were cultured and selected for homologous integration of the gene (p. 77), there would be no control over where in the chromosomes it was integrated and hence no assurance that it would be normally expressed or that it had not, on integration, disrupted the function of some other gene. These uncertainties · about integration would, indeed, also apply to somatic cell therapy, but here the risk would be spread over many separately transformed cells, some of which might be perfectly functional even if all were not. In the case of a germ cell, all the eggs are, so to speak, in one basket. An accident in attempted germ-line therapy would be present in every cell of the body of a child inheriting the damaged genome.

The human genome project

Currently much excitement is being generated by a proposal to obtain a complete sequence of DNA bases for each of the 23 kinds of human chromosome (22 pairs plus the X and Y sex chromosomes). The human genome contains approximately 3 thousand million base-pairs and its complete sequence would fill a volume of several hundred thousand closely printed pages (in practice, of course, the data would be stored in a computer memory). It has been estimated that the project would take about 15 years for its completion and would cost of the order of $3 billion – expensive even by the standards of contemporary 'big science'. In the USA an Office of Human Genome Research, with the prime function of sequencing, has been established under the National Institutes of Health, and funding is likely to be provided both by NIH and (for reasons not too evident to the outsider) by the Department of Energy. In Europe, too, substantial public funding is being made available in several countries – France, Italy and the UK – for human genome research. An international group under the title Human Genome Organization (HUGO), and with a Council composed of top-level molecular biologists and geneticists, has been set up to coordinate the effort internationally.

The most immediate benefit of this immense project would probably lie in the boost that it would give to the identification of DNA sequences of genes and of RFLPs really closely linked to them. This would enlarge the scope of antenatal diagnosis of genetic disorders.

A consequence of a complete determination of DNA sequence would be the establishment of an inventory of the amino acid sequences of all the proteins that could be made in a human cell. Genes that could encode proteins are fairly easy to recognize, and the amino acid sequences of the proteins that they code for can be deduced. Predicting the functions of proteins from their amino acid sequences is very far from easy, but it is becoming increasingly possible to make reasonable guesses on the basis of analogy with known proteins with known functions. Hence, if, for example, the metabolic function that has been lost in a particular genetic disease were known, and if the gene responsible had been mapped to a particular chromosome region, then examination of the complete DNA sequence in that region could well reveal a likely candidate for the gene being sought. Confirmation would have to come from the identification, within the candidate gene, of a molecular change that correlated completely with the potential for disease. It is important to appreciate that scrutiny of the DNA sequence from one individual will not give the answer; study of DNA variation between different individuals will also be necessary.

The human genome project has been criticized on various grounds. To many it seems unnecessary and extravagant to go for the whole genome when it seems that only a minor fraction of it comprises functional genes, with the rest looking more like padding or junk. There is a surprising number of 'pseudogenes' apparently derived from functional sequences but long since fallen into disuse and decay, and there are long tracts of DNA that have none of the earmarks of function, either past or present. Much of the apparently functionless DNA consists of rather simple and repetitive sequence. For those immediately involved, the project is likely to be akin to a virtually interminable plod through wasteland or desert, with only the occasional functional oasis to keep the interest alive.

Since there are now ways of recognizing tracts of genomic sequence that are especially rich in active genes, it can be argued that it would be far more cost-effective to concentrate on these. On the other side of this particular argument it can be pointed out that assumptions about where genes are likely to be found will almost certainly break down in some cases, so that genes would be missed. In any case, there will be a good deal of interest in the 'desert' tracts of the genome from the point of view of chromosome structure and evolution. There is no knowing what will be found in the chromosomal hinterland, and it seems almost certain that sooner or later curiosity will prevail and every region of the human genome will be explored.

Another criticism relates to the arbitrary choice of one complete human genomic sequence as the type specimen to stand for the whole species. Even if the individual(s) supplying the DNA are picked as fully functional human beings, they are sure to be harbouring personal and unrepresentative idiosyncrasies including, almost certainly, some defective genes of recessive effect that could be a source of confusion. In practice, however, such individual abnormalities are likely to amount to very little in comparison with the vast amount of essentially normal and typical sequence. Different people differ in detail in all sorts of ways, but the essential metabolic machinery and developmental programme are common to all mankind, just as are the main features of human anatomy.

There is no denying the great importance of the database that the completion of the human genome project would establish. The availability of a complete inventory of genes and their protein products would allow many short cuts to be taken in all kinds of investigations into human metabolism and development.

That said, the likely impact of a complete human genomic sequence can be over-stated. It seems not to be generally understood that there is still a vast gap between the possession of such an inventory and a full understanding of the genetic basis of human variation. To take one example of misapprehension, a science article in a relatively serious newspaper (*The Guardian*, 13.12.88) stated that 'there is a real belief among scientists that the Human Handbook will reveal the genetic secrets also of the musical ability of a future family of Beethovens or the mathematical ability of a future family of Bernoullis'. Such hyperbole overlooks a basic point of genetic methodology alluded to above. In order to attribute a particular trait of a whole organism to a particular variant of a gene, one has to analyse a considerable number of different individuals – preferably drawn from a few extensive pedigrees – in order to establish the likelihood of a causal connection.

Moreover, to make the analysis possible at all, the effect of the gene variant would have to be sufficiently pronounced to stand out above the background of variation due to other genes affecting the same character. Metabolic disorders due to single-protein deficiencies are indeed due to single-gene mutations. And it may well be that, in the long run, major effects on susceptibility to such less obviously genetic conditions as heart disease, will be attributable to particular identified gene variants. Such information will not, however, be obtained simply by perusal of the 'Human Handbook'; a large amount of careful analysis of genetic variation will be required in each and every case. When we come to complex mental traits such as general intelligence or musical talent, the possibility of identifying single relevant genes becomes much more remote.

Disentangling the genetic causes of human variation from the effects of environmental differences is, for all but the most clear-cut traits, a formidable problem in itself. But even granting a major role to the genes, there is still a vastly complex network of interactions, operating both in space and in time, intervening between primary gene action and the qualities of the whole person. One does not need to be an obscurantist to doubt whether the human DNA sequence will ever be an open book.

Fears of misuse of DNA technology in human affairs

In considering the apprehension that has undoubtedly been aroused by the prospect of human applications of DNA technology, it is necessary to distinguish between diagnosis and therapy.

Diagnosis

The ethical objection to the use of DNA probes for antenatal diagnosis is that, at least up to now, it has implied a readiness to resort to abortion if the fetus is found to carry genes that would condemn it to serious handicap. For people who, for religious reasons, would find abortion totally unacceptable, antenatal genetic screening has had no purpose – except, perhaps, to give early warning that the baby is likely to need special treatment, and this seems a rather weak reason. Phenylketonuria, for example, is routinely detected postnatally in time for remedial action and for most other genetic conditions there is no treatment in any case.

As we pointed out above (p. 103), an improvement of the efficiency of *in vitro* fertilization and embryo re-implantation could make selection of embryos before the implantation stage an alternative to selective abortion. This would no doubt be more acceptable to many mothers, but it would still not satisfy those who believe that embryos, however early, should still be treated as human beings. Selection among embryos necessarily entails sacrificing those not selected.

From the point of view of the 'pro-life' movement, antenatal diagnosis is presumably objectionable as an inducement either to abortion or to deliberate sacrifice of early embryos. For those without such religious principles, the diagnostic use of DNA technology would seem to offer only benefit – the benefit of being able to avoid defective births and (usually) have healthy children instead.

There is another point to be made here, however. Even some who are not absolutely opposed to abortion may still be worried by the possibility that, given too much information about the unborn child, some people might resort to abortion for reasons unconnected with predicted handicap. The most likely reason, a powerful one in some cultures, would be a preference for male children. Most people would probably consider aborting a fetus or discarding an embryo because of its gender far less defensible than doing so because it was destined to suffer from a crippling handicap. Apart from anything else, to give free rein to parental choice in this matter might jeopardize the approximate numerical equality of the sexes, which it is clearly desirable to maintain.

Therapy

Gene therapy is probably more widely controversial, but it is far from clear that, if applied to somatic cells, it would be any different in principle from other more familiar kinds of rescue operation such as kidney transplants. In so far as it entails ethical problems, they seem to be the same as beset 'high-tech' medicine in general. There is the problem of balancing the legitimate enthusiasm of the pioneering therapist against the right of the patient not to be pressurized into accepting unduly speculative treatments. There is also the problem of priorities in allocation of limited clinical resources – too great an emphasis on ambitious treatments of relatively uncommon conditions could take resources away from less exciting kinds of medical care for which there was a greater need. Neither of these potential sources of conflict is peculiar to operations with DNA.

Germ-line therapy is a different matter altogether. It would be intrinsically hazardous, it would probably not work, and in any case it would be largely pointless. If the objective were to avoid defective births, antenatal diagnosis combined with selective abortion is a far easier and more acceptable way of achieving it. Religions that reject abortion are hardly likely to look any more favourably on risky experiments with embryos intended for survival, which is what every attempt at germ-line therapy would amount to.

Eugenics

Behind much of the anxiety about increasing knowledge of the human genome is the fear that it might lead to compulsory eugenic measures. This appears to have been the basic reason for the extreme suspicion of the human genome research shown by the Green parties of Europe, whose view influenced the European Commission to delay giving funding to any investigation of the human genome until its social and ethical implications had been explored. This decision particularly reflected the belief of the German Green Party that knowledge of the human DNA sequence could encourage a resurgence of practical 'eugenics' of the sort used against Jews and gypsies by the Nazis.

Some of those concerned about possible applications of DNA technology to eugenics fear that it may, in the future, be possible for those in power not only to eliminate what they regard as undesirable traits, but even to restructure human beings to their own design. This is greatly to overestimate the present and foreseeable future power of genetic engineers. There are real prospects for gene therapy but, without wishing to be in any way derogatory, one can liken the aspiring gene therapist to a novice motor mechanic. He has noticed that a particular kind of malfunction tends to occur when a particular bit of wire has come loose. He has only a rough and uncertain idea of what the wire does, but he thinks that if he fixes it the malfunction may be cured. In this he is probably right, but he is far indeed from knowing how to take the whole machine apart and reassemble it, let alone from being able to design a new model.

If an authoritarian regime really wished to steer the human population in a particular genetic direction, it could do so without the enormous expense and difficulty of mass DNA manipulation. It could just follow the traditional animal breeding principle of selective breeding of the desired phenotypes. The defence against the possibility of such tyranny lies in politics, not in denying ourselves knowledge.

References

1 Jeffreys, A. J., Wilson, V. and Thein, S. L. (1985) Individual-specific fingerprints of human DNA. *Nature* **316**: 76–79.
2 Howlett, R. (1989) DNA forensics and the FBI. *Nature* **341**: 182–183.
3 Weatherall, D. J. (1985) *The New Genetics and Clinical Practice*, 2nd edn., 206 pp. Oxford University Press, Oxford.
4 Dickson, D. (1989) Early prenatal diagnosis. *Br. Med. J.* **299**: 1211–1213.
5 Coutelle, C., Williams, C., Handyside, A., Hardy, K. and Winston, R. (1989) Genetic analysis of DNA from single human oocytes: a model for preimplanation diagnosis of cystic fibrosis. *Br. Med. J.* **299**: 22–24.
6 Handyside, A. H., Kontogianni, E. H., Hardy, K. and Winston, R. M. L. (1990) Pregnancies from biopsied human preimplan-

tation embryos sexed by Y-specific DNA amplification. *Nature*
344: 768–770.

7 Marx, J. L. (1989). The cystic fibrosis gene is found. *Science* **245**:
923–925 (referring to papers by F. S. Collins and L.-C. Tsui and
their respective research teams in *Science* **245**: 1059–1065 and
1066–1080).

8 Friedmann, T. (1989) Progress toward human gene therapy.
Science **244**: 1275–1281.

9 Wu, G. Y. and Wu, C. H. (1986) Receptor-mediated gene delivery
and expression *in vivo*. *J. Biol. Chem.* **263**: 14621–14624.

9 | RISK ASSESSMENT

The previous chapters may have conveyed the impression that the hazards of genetically engineered organisms as now handled are not such as to cause serious immediate concern. This is generally correct; by itself genetic manipulation does not normally confer qualities that will make an organism more harmful to mankind or to its environment. When genetic manipulation is being done under controlled laboratory conditions, the hazards of such procedures, while naturally requiring care, have not proved to be a significant addition to those already presented by the organisms being used. Of course, when the organisms are designed to be robust in order to survive in the wild, or when they are used on a very large scale in agricultural practice, then new problems may arise. Our experience in coping with these is limited; and so when we consider the problems of regulating this new sort of scientific work, we become aware of the depth of our uncertainties about the processes and their containment. It is in such cases that the skills of risk assessment will be deployed to their fullest extent.

The precedent from genetic manipulation

In considering how to assess the risks associated with the release of genetically engineered organisms or viruses, it may be useful to recall the precedents set by the Genetic Manipulation Advisory Group (GMAG) and their successors, the Advisory Committee on Genetic Manipulation which is now an agency of the Health and Safety Executive. The system in current use has been unchanged in principle since 1979, when the GMAG adopted a method of risk assessment that had been devised by the eminent molecular biologist Sydney Brenner. Broadly speaking, the effect was to place less emphasis on the supposed inherent risk of the DNA itself and more on the scenario through which it could become an actual threat to health.

Attention was focussed on three considerations labelled access, expression and damage. The first of these headings covered the likelihood that the recombinant DNA would escape from the system of containment and gain access to the human body. The access factor was rated most highly when the DNA was cloned into a bacterium, such as wild-type *Escherichia coli*, that could readily colonize humans, lower when the cloning vehicle had no such natural ability and lowest when it had been 'crippled' by genetic modification. Expression referred to the ability of the DNA to get itself transcribed and translated into some product. The expression factor was highest when the cloning vector had been deliberately designed to promote expression of genes inserted into it and lowest when the vector was designed to preclude expression. Finally, the damage factor was a measure of the ill effect of the DNA on human health if it both gained access and was expressed. It was obviously highest where the product was known to be a toxin and lowest where it was known to be harmless.

The problem with this very logical scheme was the great difficulty of quantification of any of the three factors in any particular case. This difficulty was dealt with in a very bold and arbitrary way. Instead of attempting to make precise estimates of risk, which would certainly have been spurious in view of the great difficulty or impossibility of empirical tests, each experiment was assigned, with respect to each of the three criteria, to a very broad and approximate category. Virtual certainty that the DNA could gain access, or be expressed, or pose a danger was in each case represented by a factor of 1. Below this there were different levels of supposed improbability represented by the factors 10^{-3} (one in a thousand), 10^{-6} (one in a million) or 10^{-9} (one in a thousand million). Then these factors were multiplied together to give an overall measure of the chance of the DNA actually damaging human health. If the product came to 10^{-15} (that is one chance in one million billion) or less, the proposed experiment was judged safe under conditions of good microbiological practice – which, incidentally, is defined as something a good deal more careful than most molecular biologists had been accustomed to before genetic engineering came on the scene.

It is quite easy to scoff at this system. The thousand-fold difference in hypothetical risk between experiments placed in adjacent categories could not possibly be taken literally, and the idea that one needed to take any precautions at all when faced with estimated odds against danger of one million billion to one might seem absurd – anyone who took this order of risk seriously would hardly dare to draw a single breath. But the GMAG system did embody two important principles that it would be well to keep in mind in the new context of environmental release. The first is that the probability of an outcome that depends on a sequence of events is the product of the individual probabilities of all the steps in the sequence. The second is that large uncertainties in calculation need to be compensated for by large safety margins. The system provided a fairly rational basis for deciding that some procedures were likely to be far more, or far less, risky than others, and the built-in safety margins appeared sufficiently broad to accommodate the large uncertainties in determining the absolute magnitudes of any hazards. With hindsight, we can say that, however arbitrary it may have seemed (and still seems), it was a social and political success. It provided a formula for compromise between alarmism and complacency – a formula that allowed work to proceed safely and for experience to be gained. It was a system that led to decisions that nearly everyone involved felt were adequately cautious.

Extension to released organisms

However well it worked for laboratory experiments, the GMAG formula can hardly be applied unchanged to environmental release. The very term 'release' implies access to the environment and, although farm animals can be effectively fenced in and thus not really released, it will hardly be practicable to apply effective containment to crops growing in large open fields of bacteria inoculated into the soil of such fields. One can, however, distinguish between release of the engineered organism as such and spread of modified genetic material through hybridization with wild species. An informed estimate can be made of the chance of this spread of genes, taking into account the availability of related wild species or strains capable of receiving genes from the engineered variety by hybridization, the viability and fertility of the hybrids where these have been tested and the likelihood that the transferred novelty would confer a competitive advantage. In the case of an engineered bacterium, one would want to know whether there was any known way, for example through plasmids transmissible between different strains and species, that could effect transfer of novel genetic material. These complex considerations could all contribute to a rough estimate of a secondary access factor.

The expression factor is hardly relevant in the case of an organism engineered for environmental release; if its genetic novelty were not expressed there would be no point in releasing it. Thus, if we are considering the primary product of genetic engineering designed for environmental release, everything rests on the danger factor; in the case of hypothetical products of secondary spread the overall risk will be the product of the danger and secondary access factors.

It is hardly possible to propose even a three-orders-of-magnitude estimate of danger unless one has some hypothesis as to how danger might arise. The obvious examples of projects involving foreseeable risks are those for making new kinds of live vaccines. Assessments of risk in this area will carry more weight when the roles of particular bacterial or viral genes in determining pathogenicity are better understood. Meanwhile, regulatory agencies will have to make cautious and approximate assessments on the basis of the best evidence available. Their task will be the more difficult in that the hypothetical risks of new vaccines will have to be balanced against the prospect of meeting particularly urgent human needs.

The genetically modified farm animals or crop plants at present under consideration present a different kind of problem. It is difficult to see how any of them could be dangerous in themselves unless they have the potential to release pathogenic viruses, a possibility that can and should be avoided. The main perceived risk, which applies to crop plants and hardly at all to animals (except perhaps fish), is that some of the new types will turn out to be invasive, swamping important crops or natural habitats. What has to be assessed here is the likelihood of invasiveness, and the ecological or agronomic damage that might result from it. To some extent this question can be approached experimentally through the use of contained 'microcosms', but these can never simulate the enormous variety of the outdoor environment. Furthermore, to the extent that the danger is thought to reside in transfer of genes to wild species, it may not be easy to guess the kind of hybrid that it would be important to test.

Coping with these problems requires knowledge rather remote from the molecular biology that has dominated discussion hitherto. The scientists best qualified to judge these questions are biologists who have made a close study of the particular species groups to which the organisms under consideration belong. Knowledge of comparative morphology, taxonomic relationships, life histories and environmental preferences must be the best basis for judging how a particular genetic modification is likely to affect competitive ability or invasiveness, either of the organism primarily modified, or of possible secondary recipients of engineered DNA. In other words, we would look to taxonomists and students of genetic variation in relation to the fine detail of ecology. This is the kind of knowledge that tends to be included in the category of 'natural history', a term used nowadays, more often than not, as a term of disparagement. It attracts very little research funding or student interest, and many university biological departments seem virtually to have given up the struggle to keep it going. It seems important to reverse this trend if we are seriously concerned about environmental hazards of engineered organisms – or indeed about the many other kinds of environmental degradation that are arguably more immediate and serious.

In our view, the discussion of the risks of genetic engineering has been too much dominated by an over-dramatic view, verging at times on the apocalyptic. A comparatively moderate opinion sometimes expressed is that we are faced a very small chance of an extremely large catastrophe. This 'zero-infinity' postulate has a certain ironic intellectual appeal, the point being that the product of the two terms can take any value that one might like to imagine. Reality, we believe, is likely to be altogether more mundane. It seems more realistic to think in terms of fairly small chances (perhaps roughly calculable) of less-than-catastrophic problems that one could overcome or live with. The risks in prospect are likely to be no different in kind from those that have always been associated with the introduction of the novel products of plant and animals breeding, or new vaccines. That is not to say, of course, that they do not need to be anticipated and guarded against.

What may the unforeseen provide?

It is sometimes said that predictions of the probable harmlessness of genetically manipulated organisms are cast in doubt by the inevitability of new scientific discoveries. Of course new discoveries will be made in, say the ecological genetics of microorganisms (a subject almost called into being by this very consideration). How can we guess what effect they might have? We can perhaps gain some clues by looking backwards, at findings in the not too distant past which were unforeseen and which expanded our ideas of what is possible in the field of genetics, either molecular, microbial or ecological.

We can drive these findings into two types. The first has to do with the mechanisms that exist for the propagation and expression of the genetic material, while the second does not upset our ideas about basic mechanisms but has to do with the contingent facts of what actually happens in the real world.

The first class, of 'fundamental surprises', includes the following:

1 The existence of intervening sequences ('introns') in the genes mainly of higher eukaryotes. These are transcribed along with the coding portions ('exons') of the gene, but are spliced out of the messenger RNA before it assumes its mature, translatable form.
2 The capacity of RNA to adopt a catalytic role in biological

reactions; this was once regarded as the special province of pro-
teinaceous enzymes.
3 The role of 'reverse transcriptase', the RNA-dependent DNA
polymerase that copies RNA into DNA in the life cycle of certain
viruses.
4 The non-universality of the genetic code, especially in the
mitochondria of eukaryotes.

With hindsight, we can see that all of these extend the known
characteristics of certain classes of macromolecules, rather than
providing something wholly new. The special properties of RNA
are inherent in the 'adaptor' role of a certain type of RNA, the
amino–acyl transfer RNA, predicted in the late 1950s. The ability
of RNA to be copied into DNA and the (rather limited) non-
universality of the code are not incompatible with the complex-
ities of the biochemical events in base pairing. The likely impact
of such findings on our view of the proclivities of DNA in the
natural environment would seem to be limited.

The second class of 'contingent surprises' is very varied; a few
examples are as follows:

1 The existence of 'transposable elements', blocks of DNA that
change location in the genetic material. That their discovery is not
new (for plants it dates from the 1950s, for bacteria from the late
1960s to early 1970s) should not blind us to its originally apparently
revolutionary quality.
2 The ability of the Ti plasmid DNA of the bacterium *Agrobacterium
tumefaciens* to transfer itself into plant cells, followed by the
insertion of a segment ('T-DNA') of this plasmid to enter plant
chromosomal DNA.
3 More recently discovered is the ability of some bacterial plasmids
to move between widely diverse bacteria (the so-called 'Gram-
positive' and 'Gram-negative' types) and even to yeast [1].
4 Similarly, we have recently become aware of the presence of
unexpectedly large numbers of bacterial viruses in natural bodies
of water. These must exert strong selective pressure on the bacteria, and may in addition mediate DNA transfer between them
(by the process called 'transduction') [2].

Unlike the class of 'fundamental surprises', the 'contingent
surprises' certainly have implications for our theme. But these
implications can work two ways. We can point to the fact that

since these observations involve the remarkable *natural* move-ment of genetic material, within cells and between them, they suggest that the diversity of natural genetic transfer is so vast that nothing we can do is likely to make much difference. Or we can look at the awesome amplifying power of such systems, to spread throughout the natural world any construct made by our *in vitro* methodology that is allowed to escape. The dichotomy in view-points seems inescapable. Thus the problems of the risks of released genetically manipulated organisms cannot be definitely solved along the same lines as those of the classic experimental and field sciences.

References

1 Stachel, S. E. and Zambryski, P. C. (1989) Generic trans-kingdom sex? *Nature* **340**, 190–191.
2 Sherr, E. B. (1989) And now, small is plentiful. *Nature* **340**, 429.

10 | MANAGING THE UNCERTAINTIES OF RISK ASSESSMENT

Special problems of biohazards

Biohazards present significantly greater difficulties of assessment than hazards from physical plant or even natural accidents. We may appreciate the reason for this difference in terms of the idea of information. Man-made machines and living organisms both carry information about their own functions. The unique feature of organisms, however, is that their information is self-propagating and so can produce more organisms of the same kind. Moreover, as we now know, it cannot be assumed that the information will always remain within the bounds of a single kind of organism. Especially in bacteria, informational pieces of DNA can be transferred between species at low but significant frequencies through the agency of plasmids or viruses, or even without the help of such vectors. Some of the species in receipt of information could be harmful to us. The microbial flora of natural and agricultural environments is so diverse that there is little hope of enumerating and quantifying all the possible pathways of information transfer; and it is even quite difficult to monitor empirically any particular DNA fragment after it has been released. Most probably it disappears, but who can be sure that it is not hiding out somewhere ready for a resurgence when opportunity arises?

Here we are very far from the controlled laboratory experiment upon which 'hard' science is traditionally based. In an inevitably poorly defined situation it is very easy to be guided by ill-founded or even self-interested perceptions. If debate over the hazards of genetic engineering, and over planned release in particular, is to have any scientific content, we must be clear about the crucial differences between the style of laboratory science and that involved in controlling hazards.

Almost by definition, a laboratory experiment cannot be conducted in an ill-defined situation. Even if it is exploratory, or motivated by speculative theories and ideas, in its specification and techniques it must be precise and controlled, or else it is valueless. Normally, there will be an hypothesis being tested; and that must be stated as clearly as possible. At the end of a successful experiment there will be a few well-defined alternative interpretations and conclusions, which can be presented for review and criticism.

All this is possible when the substances or organisms in the system under investigation are well defined and accurately reproducible, and can be subjected to controlled variation taking one variable at a time. In the case of environmental hazards, very few of these conditions will hold. Although there have been few cases of damage to the general public arising from biological technology, the problem is a familiar one with chemical and radioactive agents. Sometimes there is a situation approaching a local epidemic, and it is only selfishness and stubborness of the offending institutions that prevents a quick and clear identification of the cause; such was the case with the incident at Mishimata in Japan, where a discharge of mercury into the sea resulted in the poisoning of people who ate locally caught fish. But more frequently, the evidence of the existence of a problem is itself debatable; and the assignment of a cause becomes very difficult indeed. The techniques of statistical epidemiology must be applied; and in difficult cases these may be as controversial among practitioners as they are obscure to the public. The case of the 'leukaemia clusters' found in the vicinity of some nuclear power plants (but not only in such places) is a case in point.

Understanding public concern

The specification of possible future hazards from an untried technology is a still more uncertain task, whose scientific basis is very thin indeed. The skills and presuppositions that make for success in the laboratory, the concentration on testable hypotheses and the dismissal of vague theories, can sometimes become counterproductive. A fine balance of attitudes becomes necessary; while it is impossible for scientists to chase down every scare story about disease or pollution, there have been important cases when evidence that was 'merely anecdotal' eventually provided clues to a serious problem. One such was 'Lyme Disease' in Connecticut, where an epidemic of rheumatic fever among children became a matter of public alarm while still unrecognized by the medical profession; it turned out to be due to a spirochaete parasite transmitted from deer by ticks. Also, the attitude of demanding scientific rigour in an argument about a suspected hazard in order that it be taken at all seriously, may be misunderstood by a worried public as representing a lack of concern. For the public's worries may on occasion be ill-formed and perhaps also ill-informed; but they are not necessarily silly or irrational because of that. A concentration of focus on the hazards that can be clearly specified, and a dismissal of the others, can be interpreted as scientific hubris of the sort that has on occasion produced unacceptable environmental damage. Yet to expect scientists to go through the motions of assessing every conceivable hazards, however speculative or remote, is to misunderstand the workings of science. The need is for a balance, in which scientific uncertainties are managed by prudence and tact, and in which the differing perceptions of the different parties are recognized and respected.

It is likely that much of the acrimony that afflicted the first American debate over the hazards of recombinant-DNA research was caused by this difference of perceptions. The scientists had developed techniques whereby they had solved a host of new and challenging problems, and they saw no reason why it would be different for any of those to be encountered in the safe operation of these techniques. Their critics felt otherwise; they had been told of scientific problems of a novel environmental and epidemiological character, which the scientists themselves had deemed sufficiently serious to justify a research moratorium. With their heightened concern for ethical and theological issues, mixed with traditional town-gown politics, they felt justified in withholding their consent to the scientists' reassurances. If we are to get beyond that sort of mutual incomprehension, which could (in the present political climate) have disastrous effects on the development of biotechnology as a force for progress, we will need to think clearly about the management of uncertainty in this most difficult area.

The first step is for all sides to appreciate that, in this area, there can be none of the certainties that have traditionally been associated with scientific knowledge as it has been taught and popularized. For the management of uncertainties in this new area of science, so different from traditional laboratory work, we could do well to consider principles and techniques developed in other fields of practice. The first is a very general one, suggesting an enrichment of what is already enshrined in the practice of science. This is the maintenance of a critical presence, so that the quality of the work is ensured. In science this is accomplished through peer review, refereeing, and public discussions at seminars and public meetings. When scrutiny is focussed on research projects and results, as in the case of science, it is inevitable and proper that only those with a trained technical competence should be part of the process. But when, as in the present case, the questions are of what might somehow happen, in a variety of areas of human experience (touching on issues of social affairs and ethics) and with greater or lesser degrees of probability or plausibility, the relevant competence is neither so tightly defined nor so exclusive in its scope.

In such circumstances, the presence of lay persons on regulatory agencies can be strongly defended. Professional scientists may be over-influenced by the intellectual excitement of successful technology and over-confident about the degree to which they are in control. And even where the scientists do have the necessary objectivity and detachment it would not be reasonable to expect the public to take this on trust. Lay representatives should not, of course, be selected for their lack of knowledge; it is clearly desirable that they should be able to understand what the scientists are talking about. It is only necessary that their first concern should be seen to be with human risks and benefits rather than with the advance of knowledge as such.

The problems of regulating hazards are in some ways similar to those encountered in the ordinary courts of law, where decisions must be made under conditions of uncertainty, and where scientific information, however certain in the abstract, is sifted like any other sort of evidence. The jury is not expected to deliver its verdict as if it were an indubitable truth; an irreducible element of doubt is explicitly allowed for. But then this doubt must be operated in relation to a 'burden of proof', so that (in our tradition) the accused is deemed innocent until (effectively) proved guilty. Such a principle is not foreign to scientific practice; whenever a statistical test is made, in relation to a particular level of a 'confidence limit', a relative evaluation of the costs of the two possible errors (false-positives and false-negatives) is being implicitly invoked. When in a debate on regulatory policy the perspectives of laboratory-trained scientists and environmental activists are in direct conflict, this may be understood in terms of the implicit burden of proof which the two sides are applying to their uncertainties. Put most simply, is the process deemed safe until shown to be dangerous, or vice-versa? Where the issue of differing burdens of proof to be made clear and explicit, then at least there would be some awareness of why conflict was occurring, rather than the mutual incomprehension that so frequently occurs in debates on risk and pollution issues.

Further, without attempting to copy in any way the management of uncertainty in the judicial process, we may adapt another of its principles, in relation to the complexity of evidence. Working scientists know that the raw data from their equipment do not automatically constitute 'facts'; these must be worked up (as by statistics) and then interpreted theoretically, before they can function as evidence in an argument leading to a conclusion, which is intended to be of sufficient reliability to count as a 'fact'. All this is accomplished largely informally, using techniques and inferences that are established by a working consensus of the relevant colleague community.

In the case of hazards, where the underlying data may be sparse and ambiguous, and no evidence can be conclusive, certain disciplined techniques can be of great assistance. These operate qualitatively in the elicitation of possibilities, complementary to the statistical techniques that lead to quantitative estimates of likelihood. Those concerned with the regulation of the new biotechnology have already begun to experiment with hazard-assessment techniques designed especially for this work. In the UK, development is in progress on a technique known as GENHAZ. This consists of a structured procedure for assessing risks which records the possible consequences of the widest possible range of unintended eventualities [1]. While this technique is extremely useful for calling attention to possibilities that might otherwise have been overlooked, it cannot by itself evaluate the quality of the information on which a particular scenario is based. Hence there are opportunities for much useful work on these techniques of making best use of the information that is available, and converting it into effective evidence in a hazard assessment.

Coping with the scarcity of information

One crucial problem is the paucity of relevant information. The fields of taxonomy and microbial ecology have not been in the mainstream of biological research, and the lion's share of funding has been going into work at the molecular or cell level. Indeed, there is now official recognition in the USA that systematics (now enjoying only about 0.3% of the funding of biomedical research) is in danger of going into rapid decline as its ageing experts retire with too few trained scientists to replace them [2]. Yet it is only through these relatively neglected fields that we will solve the problems that are created by the successful applications of the popular ones.

Until that situation is redressed, and it will require a long lead time for the expansion of teaching in those fields, those who assess hazards will do so in the absence of information that in some contexts might be considered crucial. Of course, it is in such situations that burdens of proof must operate, if not explicitly then implicitly, and where personal prejudgements can become decisive. A useful task would be to devise means for the disciplined management of this sort of uncertainty.

Although we cannot embark on this here, we might offer a brief sketch of some of the steps in such a procedure. First, when some requisite information turns out to be not available, we may enquire why not. Although it is not always easy to find out about such things, it might be possible to discover whether someone had intended to do it but funding had been refused. The reasons for that could be a combination of uncertain feasibility, high cost or low priority. Thus if the necessary methodology of a project is inherently very difficult, funding agencies would require proof of very high priority for resources to be put into it.

Obtaining a profile of possible research that has *not* been done, a map of our ignorance as it were, could be an illuminating exercise in itself and could provide guidance for changed priorities. Secondly, perhaps the research had been done, but the results are confidential to some commercial firm or government agency. In some cases, as with defence-related biological research, deep secrecy is accepted as essential for national security; but otherwise, a regulatory programme is severely hampered if the very existence of confidential research results is itself a confidential item. Finally, given the absence of the required information, there could be an estimation of what difference would be made to a regulatory decision if it were available. This last procedure, aggregated over a number of cases, would help to focus attention on those questions on which further research is most urgently needed.

All this may appear to some to be an overly elaborate solution to what is, after all, a problem in scientific common sense. But the planned-release techniques raise rather special problems, so that (within the limits of feasibility) more conscious care is required in the analysis of its hazards. For example, a planned-release technique may involve an organism that is already dead or intended to expire after its work is completed. In that case, the task of assessment is simpler, for any possible harmful effects must be transferred across the barrier (not absolute, to be sure) of the short life and perhaps moribund condition of the organism. But sometimes the organism is intended to survive, at least for some time, in the wild. The question of regulatory principle is then whether it should be guaranteed not to displace any of the existing related biota, or (less restrictive) merely not to cause disturbance to the environment. In the former case, an exercise in biological fine tuning must be involved, so that the organism will be just strong enough but not too strong. There must also be a further exercise, in proving adequately (keeping in mind what we have already said about burden of proof) that the engineered organism will indeed have the desired property. Such studies could indeed strain our resources of microbial ecology in many cases; and yet if containment in the appropriate degree is to be guaranteed, it is hard to envisage any alternative.

We believe that with cases such as these, there will be many decisions where the scientific component is far from conclusive, and so where the skill and integrity of those making the decisions, as perceived by the public, will be crucial. It is for this reason that we advocate as much clarity and publicity as possible in the management of uncertainty in the control of the hazards of genetically engineered microorganisms.

Principles of ecological assessment

The hazards of genetically engineered organisms result from unplanned interactions between the organism and its environment, which may be good for it but bad for us. To control these interactions, we need to know about the organism and its genetic materials, and also about the relevant aspects of ecology. As we have already observed, in biology the study of the environment lags far behind the study of the organism. The field sciences have lacked scientific excitement for attracting researchers; they are in ways like a relic of the gentlemen-amateur science of the Victorian age and before. The crisis is now attracting notice, and it is becoming appreciated that unless present trends are reversed, this imbalance will increase, perhaps eventually approaching the point where there is a drastic shortage of personnel with the field-science skills necessary for assessing the hazards that the laboratory sciences are creating.

Even if we begin to redress this imbalance, there will always remain severe uncertainties in the predictive aspects of ecology. Some argue that the natural environment is so complex, involving interactions so numerous, so varied and so interrelated, that a 'predictive ecology', on the analogy of physics, chemistry or molecular biology, is impossible. Others argue that it *must* and that (perhaps therefore) somehow it *can* be done. Whatever may be the eventual outcome of that debate, we need now to be able to make reasonable predictions to guide our regulatory decisions. By drawing on past experience, it is possible to draw up guidelines, focussing attention on those aspects of an organism in its environment that are crucial for assessment.

1 *Specificity*. As with conventionally produced organisms used in pest control, genetically engineered organisms should be screened against a wide range of possible targets and released only after their specificity has been established.

2 *Predictability*. Although unexpected effects have not been common following natural invasions, every effort ought to be made to reduce the possibility of unforeseen consequences following a deliberate release, to monitor for their occurrence and to contain them when they occur.

3 *Reproduction and spread*. As demonstrated by many weeds, high rates of both reproduction and dispersion are likely to mean difficulties in controlling a released organism.

4 *Scale of use*. Success in establishing a novel species will depend to some extent on the scale and frequency with which it is disseminated in the environment.

5 *Reversibility*. Some past invasions (such as the North American muskrat in Britain) have been reversible. Where, as with microbes, this is unlikely to be achievable, even greater reliance must be placed on a thorough assessment of an organism's range and specificity.

Even with such principles in force, there can still be no guarantee that any particular introduction will be totally benign. We should not forget, nor indeed will scientists be allowed to forget, that CFCs were brought into widespread use only after the best scientific advice was that these simple chemical compounds are truly inert and could therefore cause no environmental harm. Such cautions are even more appropriate when organisms are considered for large-scale routine commercial use; for then all the above criteria present very different problems. This difference will lead to a sixth ecological principle, of quality assurance in application; and we should argue for this in some detail.

Earlier we mentioned the difference between the conditions of traditional laboratory research and those of the control of hazards. There is an analogous difference between natural hazards and those that are crucially dependent on industrial practice. This difference erupted in a sharp debate on the regulation of biotechnology, and so it is particularly relevant to our concerns. The case, as described by Brian Wynne [3], was the examination, by the 'Lamming Committee' on behalf of the European Commission of five hormones for use in the beef industry. Two, which were fully synthetic hormones, were deferred for further study, while three 'nature-identicals' were judged acceptable under certain conditions of use. The Commission's own experts interpreted this as scientific approval, and on that basis were prepared to permit their use. In that way they would both support the beef industry and also encourage biotechnology. However, there ensued some intense political lobbying, with the result that the Council of Ministers rejected the experts' advice and banned the hormones.

The Commission's experts were indignant at this incursion of what one called a 'wave of ill-informed emotive popular superstition'. However, the nature of the problem is revealed in the conditions laid down by the Lamming Committee. These included: specified dose limits; non-edible site of injection; full veterinary supervision; and a minimum waiting period before sale of treated meat. In an ideal world, good farming practice would of course ensure that such conditions were met. But the real world is not ideal, and it is hardly irrational to query whether in the absence of close policing in all the agricultural districts of the Community, such standards of quality could be expected to hold. In the absence of such quality assurance on farming practice, the hazards of the hormones in question could not be assumed to be negligible.

Those who are familiar with technology and industry are keenly aware that quality assurance cannot be taken for granted; it is well known that the study and mastery of quality assurance by the Japanese some years ago was an essential part of their rise to dominance in so many fields. In the industrial context, quality relates mainly to reliability in performance; safety of products for consumers is generally assumed to be an unquestioned priority. But when hazardous substances are involved, in the context of industry or agriculture, quality will relate to risks as much as to reliability. And since it is employees, livestock or the natural environment that may be at risk, rather than consumers, the assurance of quality may become quite problematic. Hence for these circumstances, we suggest that the preceding ecological principles be supplement by another:

6 *Quality assurance in application.* The realities of practice, and of regulatory processes as well, must be included in any assessment of hazards in any context of agricultural or commercial application.

We recognize that it is not always easy to argue on the basis of imperfections in government practices, especially when such imperfections are not officially admitted. But to assume that quality assurance will automatically be maintained on all sides, is to invite the sorts of damaging incidents that have brought the nuclear power industry down from its early bright promise to its present uncertain future.

Maintaining public confidence

The function of the above principles is two-fold: to ensure that safety in genetic engineering is maintained, and also to ensure that it is seen to be maintained. We have already commented on the tendency to optimism among the laboratory-based scientists, that manifested in the debates on recombinant DNA research in the 1970s. It is now more than 20 years since a public awareness of environmental problems became an issue for the conduct of industry and technological innovation. The confusions and extremism associated with the early struggles have abated, as both sides have matured in their understanding of the problems. However, public opposition can still be a potent force; now leading German manufacturers are building biotechnology research centres in the USA rather than at home [4].

When the relevant public is really aroused, proposers and regulators face a most difficult situation. It is scientifically impossible to prove the impossibility of an unwanted event; and this may seem to be what is demanded by protestors. Equally, it is politically impossible to prove an 'acceptability' of a risk, when a public sees it being imposed on them by an organization that lacks human concern and perhaps also technical competence.

Hence the new politics of NIMBY ('Not In My Back Yard') is an importance factor in any technological or industrial development.

It is clear that a sober and responsible mood now prevails among those responsible for the current work of managing the risks of genetic technologies. It is therefore possible for there to arise a new and mutually beneficial dialogue between the scientific and regulatory experts on the one hand, and representatives of special concerned groups and of the general public interest on the other. It is now understood that people generally do not attempt to make a quantitative assessment of the risks being imposed on them, but rather they assess those persons in authority who either create or regulate them. And in this they are not being irrational; just like a lay jury (and a lay judge) in our judicial system, they evaluate evidence to a great extent from the quality of the testimony of witnesses. Of course they can be mistaken; but the policy is prudent and effective both in practice and in principle. To continue the juridicial analogy, all it needs is for safety to be respected and to be seen to be respected, and the problem of public confidence can be resolved.

A genuine policy of this sort brings its own benefits for the work of regulation and control. For if those in charge see it merely as a cosmetic exercise, placating or appeasing an ignorant public, then the work will eventually deteriorate. The prevention of accidents of any sort requires continuing commitment and morale at all levels, especially from the top-down. Our culture and personal attitudes are not tuned to the sort of success that is measured by nothing happening. The best planned system of safety will atrophy and decay, without regular injections of commitment. The natural consequence of a good safety recorded is complacency. Equally naturally this would be followed by corruption, as realized in the concealment of incidents that might spoil the good record. If those in charge of a safety programme believe that their scientific knowledge or personal integrity makes their system immune from danger, then the consequences of their attitude will manifest sooner or later, causing harm to all concerned.

What is needed, then, is a certain attitude of scientific humility, an awareness of partial ignorance and of the human fallibility of those operating the safety system. With this would go an appreciation of the role of independent observers and even critics, in keeping the work up to a good standard. Conversely, people who are experienced in public affairs would from their side recognize the genuinely good intentions, and the real special competences, of those managing and regulating the system. There could then be a constructive dialogue (not necessarily always smooth) rather than a destructive polarized debate. There are precedents for this approach from the 1970s. One was in Cambridge, Massachusetts, after the shouting died down; a committee of citizens representing a cross-section of the population reviewed the problems at Harvard University, and eventually accepted regulations which were only a slight modification of those originally proposed. The other was the original Genetic Manipulation Advisory Group in Britain, where representatives of employees and of the public interest had equal members with the scientists, and (behind closed though leaky doors) engaged in a very effective common dialogue on the construction and operation of the regulatory system.

The effectiveness of a proper attitude is confirmed by recent experience of planned releases. Where public concerns have been respected, as in this country, there have been no problems; elsewhere, the record is mixed. But again, the respect for the public must be genuine, or else sooner or later the deception will become patent, with harmful consequences for all. As we have seen, the management of biohazards is a very different task from the scientific activity that created them. There the work has been successful in creating 'public knowledge' through research on chosen laboratory problems; here the task is coping with the inherent uncertainties of environmental problems that are thrust upon us by the very success of our laboratory researches. As we have shown in this report, our knowledge of the hazards and how to control them, is advancing along with our knowledge of the techniques of genetic engineering. This is a new sort of science, equally challenging; and on its success will depend the progress of biotechnology and all the advances in human welfare that depend on it.

References

1 Suckling, C. W. (1989) *GENHAZ – An Attempt to Apply HAZOP to the Identification of Hazards in the Release of Genetically Engineered Organisms*. Royal Commission on Environmental Pollution, 31 May 1989.
2 Nash, S. (1989) The plight of systematists: are they an endangered species? *Scientist* (16 October).
3 Wynne, B. (1989) Building public concern into risk management. (ed. J. Brown). *Environmental Threats* (ed. J. Brown). Belhaven Press, London and New York.
4 Zell, R. (1989) History feeds German fears on gene technology. *New Sci.* (26 August), 26–28.

GLOSSARY

Agrobacterium tumefaciens. The bacterial species responsible for crown gall disease of plants. *See also* **Ti plasmid**.

Allele. One of a number of alternative variants of a **gene**.

Amino acid. A component of proteins. Twenty different kinds of amino acid are joined in long chains (polypeptide chains) which are folded in specific ways to form proteins. The specific properties of each protein are determined by the sequence of amino acids in its polypeptide chain(s).

Amniocentesis. The operation of withdrawing a small quantity of the amniotic fluid surrounding a fetus so as to obtain cells for diagnostic purposes.

Amphidiploid. An interspecific or intervarietal hybrid, usually a plant, with a double (diploid) set of chromosomes from each parent. Otherwise called allotetraploids, they are generally fertile, in contrast to hybrids that have only a single set of chromosomes from each parent, which are commonly sterile.

Antibiotic. Originally a compound made by one microorganism that would kill or severely inhibit other microorganisms, usually bacteria. The term has been extended to include antibacterial drugs modelled on natural products but modified by chemistry or genetic engineering.

Antibody. A protein made by the immune system that binds specifically and with great affinity to an alien protein or other **antigen**.

Antigen. A substance, usually a protein, that will provoke antibody formation when it is introduced into an animal in which it is not normally present.

Bacillus thuringiensis. A bacterial species that synthesizes a potent toxin active against insects. Different strains of the bacterium synthesize different forms of the toxin specific for different insect species.

Bacteria. Organisms (**prokaryotes**) consisting of single cells and characterized by their comparatively simple cell structure and, in particular, by having their essential genetic DNA in the form of a single enormous closed-loop molecule.

Bacteriophage. A virus infecting a bacterium.

Base. A chemical compound able to neutralize acids. Four kinds of base – adenine, guanine, cytosine and thymine – are essential components of DNA; RNA contains the first three of these but, instead of thymine, has another base, uracil, with very similar properties.

Base-pair. A complementary pair of bases at the interface between the two strands of a DNA double-helix. *See* **Complementarity**.

Cell. The smallest unit of life, propagating by division. Each cell normally contains a full set of genetic material in the form of DNA, enclosed (except in bacteria) in a cell nucleus which divides as the cell divides. Higher organisms each consist of large numbers of interdependent cells, but these can often be induced to grow independently as free cells on suitable nutrient culture medium.

Cell fusion. Occurs naturally in sexual reproduction, with the union of female and male germ cells, but can be induced to occur between other (somatic) cells in culture.

Chromosome. A structure in the cell nucleus containing DNA; replicating at nuclear division so that one copy passes into each daughter nucleus. **Prokaryotes** (bacteria) do not have cell nuclei, and have a single DNA loop in each cell that is usually also called a chromosome. Organisms other than bacteria (**eukaryotes**) characteristically have their nuclear DNA divided between a number of chromosomes in their cell nuclei, the number depending on the species. Sexual (germ) cells generally contain a single (**haploid**) set of chromosomes. The body (**somatic**) cells of higher organisms each contain a double (**diploid**) set, one set from each parent.

Clone. A population of organisms, cells or molecules replicated from a single progenitor, and hence identical by descent.

Cloning of DNA. The selective replication of a particular piece of DNA by sealing (ligating) it into the autonomously replicating DNA of a bacterial plasmid or virus (*see* **Vector**) and introducing it into a cell (usually a bacterium).

Code (for amino acid sequences in proteins). The mechanism whereby the sequence of bases in **messenger RNA** (itself a transcript of a gene) specifies the sequence of amino acids in the protein determined by the gene (*see* **Codon**).

Codon. A sequence of three bases coding either for an **amino acid** in a polypeptide chain of a protein, or for chain termination; 61 of the 64 possible codons each code for one or other of the 20 kinds of amino acid in proteins (one also signals chain initiation), and the other three are termination signals.

Commensal. An organism, usually a bacterium, that lives within another organism, usually an animal, without causing disease.

Complementarity (of bases). The property of specific pairing, by reason of mutual fit, between adenine and thymine and between guanine and thymine in the double-stranded structure of DNA. Complementary pairing can also occur between adenine and uracil in RNA.

Complementary RNA (cDNA). A DNA strand synthesized on an RNA template through the catalytic activity of the enzyme reverse transcriptase (*see* **Retrovirus**). The sequence of cDNA reflects, through the rules of complementary pairing of bases, the sequence of the RNA strand on which it was synthesized.

Containment. Conditions for manipulation of organisms such that their escape into the general environment is ruled out with a greater or lesser degree of assurance depending on the containment category.

Diploid. Having a double set of chromosomes in each cell nucleus. The somatic cells of most higher organisms (animals and seed plants) are diploid.

DNA (deoxyribonucleic acid). A class of large molecule consisting of long chains of **nucleotides**, each nucleotide comprising a base linked to a special sugar (deoxyribose) linked to a phosphate molecule. The DNA chain is held together by phospate–sugar–phosphate bonds and the bases project laterally from the main chain. Native DNA is predominantly double-stranded, with paired chains held together by complementary pairing of bases (*see* **Complementarity**).

Dominance. Describes the situation in which the activity of a gene on one member of a pair of chromosomes in a diploid cell masks the effect of a different form (**allele**) of the same gene on the other member of the pair. Normally functioning ('wild-type') genes are usually (not always) dominant to their defective mutant derivatives. *See also* **Recessivity**.

Double helix. The structure of double-stranded DNA, with each strand wound around the other in a right-handed sense; there are normally about 10 base-pairs per complete turn of the helix.

Downstream. Refers to position in DNA with reference to the direction of transcription into RNA or of translation of RNA into polypeptide (the two directions are the same within any particular gene). *See also* **Upstream**.

Electroporation. Treatment of cells with pulses of electricity to promote their uptake of DNA for genetic transformation.

Endosperm. The nutritive tissue surrounding the embryo within a plant seed.

Enhancer. A DNA segment with the property of enhancing the level of transcription of a gene. Enhancers may be either **upstream** or **downstream** of the genes on which they act.

Enzyme. A protein that acts as a catalyst of a particular chemical reaction.

Escherichia coli. A common species of gut bacterium, usually non-pathogenic. Harmless strains of *E. coli* are used for most gene cloning.

Eukaryote. Organism having the major part of their DNA divided between a number of chromosomes in their cell nuclei. *See also* **Chromosome**.

Gene. A segment of DNA that is transcribed into an RNA molecule to serve some particular function in the organism – often to encode a **polypeptide** chain of an enzyme or some other kind of protein.

Gene bank. A collection of DNA clones including most or all of the genes of a species.

Genome. A complete set of genetic information; the complete **haploid** set of chromosomes.

Genotype. The genetic constitution of an organism, usually revealed by breeding tests or occasionally by direct observation of the DNA.

Germ cells. Cells that unite in pairs to found each new generation in a sexually reproducing organism. In animals and humans these are egg cells from the female parent and sperm cells from the male; in flowering plants the equivalents are eggs and pollen tubes. *See also* **Haploid**.

Haploid. With a single set of chromosomes; the usual condition of the sexual germ cells. *See also* **Diploid**.

Heterozygote, Heterozygous (*adj.*). A diploid organism with two differing forms (**alleles**) of a particular gene or chromosome.

Homology, Homologous (adj.). Similarity due to relationship by descent. Thus a normal gene and its mutant derivative, or two chromosomes with the same array of genes but possibly differing in detail by mutation, are said to be homologous.

Homozygote, Homozygous (*adj.*). A diploid with two identical forms (**alleles**) of a gene.

Integration of DNA. The sealing of DNA fragments into the genomic DNA of the organism. Most living cells are able to do this.

Intron. A DNA segment within a gene but with no coding function; the segments of RNA transcripts corresponding to introns are removed (spliced out) during the maturation of messenger RNA. *See also* **Splicing**.

In vitro. Literally, 'in glass'. As when a part of a living process is made to proceed in an artificial system (e.g. test-tube reaction), in contrast to **in vivo** – in the whole living organism.

Kanamycin. An antibiotic that will inhibit growth of bacterial cells unless they have a gene that confers resistance. A related compound, code-named G418, inhibits growth of animal or human cells unless they have been made resistant by transformation with the bacterial kanamycin-resistance gene.

Ligase. An enzyme, usually of bacterial or bacteriophage origin, used to seal together (ligate) the cut ends of double-stranded DNA molecules.

Linkage. The tendency of pairs or groups of genes to be inherited together through cycles of sexual reproduction, due to their being carried on the same chromosome. All genes of a particular organism can be classified into linkage groups, corresponding to different chromosomes. Linkage is seldom complete because pairs of homologous chromosomes exchange segments (**recombination**) during **meiosis** prior to germ-cell formation. The closer together the genes on the chromosome, the less the likelihood of their being separated by recombination, and hence the tighter the linkage.

Marker (genetic). A distinguishing feature that can be used to identify a particular part of a particular chromosome.

Meiosis. The special mode of nuclear division that results in the reduction of the **diploid** to the **haploid** state. Essentially it consists of two successive divisions of the nucleus with only a single replication of the chromosomes. As a consequence, **homologous** chromosomes are segregated into different haploid products. Occurs immediately prior to germ-cell formation in animals and higher plants.

Messenger RNA (mRNA). A large class of gene transcripts that encode the amino acid sequences of proteins. Each protein-specifying gene is transcribed into one mRNA molecule which in turn is translated into one polypeptide chain of a protein. *See also* **Transcription**.

Minisatellite. The name given to a repetitive DNA sequence occurring in tandemly arranged clusters at many points in the human genome. The number of copies in a particular cluster is quite variable from one line of descent to another, and so the length of a DNA fragment containing a cluster can be used as a genetic marker (*see* **RFLP**).

Mutagen. A chemical or kind of radiation (e.g. ultraviolet, X-rays) that will induce mutations.

Mutation. A sudden hereditary change. Often confined to a single gene, when it can consist of as little as one **base-pair** replacement. But larger changes such as extensive deletions of DNA segments can also occur. Mutations occur spontaneously but their frequency is greatly increased by mutagens.

Mycorrhiza. The mutually beneficial association between plant roots and certain soil fungi.

Nucleoside. The base–sugar component of a **nucleotide**.

Nucleotide. The base–sugar–phosphate (nucleoside-phosphate) unit of nucleic acid structure.

Oncogene. A type of gene, originally found in cancer-inducing viruses, that is conducive to malignancy. Oncogenes are often mutant derivations of normal genes that, properly regulated, play essential roles in the cell.

Phenotype. All the characteristics of an organism that can be determined without genetic tests. In contradistinction to **genotype**.

Plasmid. A class of auxiliary DNA molecule, usually in the form of a closed loop, found in bacteria of many kinds and also in some yeasts. Plasmids are able to replicate independently of the **chromosome**(s), and are often infectious from cell to cell. They may contain genes of occasional use to their host organisms, e.g. by determining **antibiotic** resistance in **bacteria**. *See also* **Ti plasmid**.

Polymerase chain reaction (PCR). A procedure whereby a particular sequence of DNA, usually of a few hundred to a few thousand nucleotide-pairs, can be amplified from a minute sample to a quantity that can be analysed or cloned. It consists essentially of cycles of replication catalysed by DNA polymerase **in vitro**.

Polymorphism. The existence of alternative forms of a gene or chromosome within a population, with no one form in a large majority.

Polypeptide. A molecule consisting of a chain of **amino acids** condensed together. **Proteins** consist of one or a few polypeptide chains, each usually of the order of a few hundred amino acids on length. Each kind of polypeptide is encoded in a single **gene**.

Polyploid. An organism with some multiple of the normal chromosome number, e.g. a tetraploid has four sets of chromosomes instead of the usual two. *See also* **Amphidiploid**.

Probe. In molecular biology this usually means a specific single-stranded **DNA** sequence labelled with radioactivity (usually phosphorus-32) to aid the detection of a gene with complementary sequence to which the probe can bind.

Probiotic. A bacterial species or mixture of species which will assist the growth of an animal, presumably by competing in the gut with harmful species.

Prokaryote. Organism having no cell nuclei, but with a single DNA loop in each cell. *See also* **Bacteria, Chromosome**.

Promoter. A short DNA sequence **upstream** of the coding sequence of a gene serving as a signal for the start of **transcription** of the gene into RNA.

Protein. The type of large molecule, consisting of **polypeptide** chains in turn composed of **amino acid** units (sometimes with the addition of a sugar component – glycoproteins), that makes up much of the bulk of the cell and is responsible for most cell functions. **Enzymes**, the catalysts of metabolism, are all specific kinds of protein.

Recessivity. The lack of effect of certain gene variants when accompanied in the same diploid by a dominant form of the same gene. Most (not all) defective mutant genes are recessive to their normal functional counterparts.

Recombinant. A product of recombination.

Recombination (crossing-over). The natural process whereby homologous segments are exchanged between chromosomes, particularly during **meiosis** in sexually reproducing organisms. It explains why genes carried on the same chromosome are not completely linked – *see* **Linkage**. Recombination can also mean the artificial fabrication of a new DNA molecule by cutting and ligation (*see* **Ligase**).

Replication (of DNA). New synthesis by copying from pre-existing DNA. The two mutually complementary strands of double-stranded DNA are separated and a new complementary strand is synthesized alongside each one according to the rules of base **complementarity**.

Restriction endonucleases (restriction enzymes). Enzymes, of many different kinds from a wide variety of bacteria, that cleave double-stranded DNA at particular sequences of base-pairs – usually a palindromic sequence of four or six. A restriction enzyme will cut any particular piece of DNA into fragments usually averaging a few thousand or a few hundred base-pairs depending on whether the recognition sequence is six or four **base-pairs** long.

Restriction fragment. A product of cutting DNA with a restriction enzyme.

Restriction site. A particular sequence of bases in DNA (usually four or six) that is cut by a restriction enzyme.

Retrovirus. The type of virus that infects its host as RNA but which encodes an enzyme, reverse transcriptase, that can reverse-transcribe the RNA into DNA which is then integrated into the host genome, where it can be replicated along with the genome and transcribed back into RNA to make further virus.

Reverse transcriptase. *See* **Retrovirus**.

RFLPs (restriction fragment length polymorphisms). Differences in restriction fragment lengths yielded by homologous DNA sequences in different members of a population. Due to mutations with the effect of creating or abolishing restriction sites, or to **minisatellite** variation.

Rhizobium. A genus of bacteria that form and inhabit the root nodules of leguminous plants, conferring the property of nitrogen fixation.

Ribosome. A submicroscopic particle composed of a set of specific protein and RNA molecules, with the function of tracking along **messenger RNA** molecules and, as it goes, synthesizing **polypeptide** chain with the amino acid sequence specified in the mRNA code. *See also* **Code**.

RNA (ribonucleic acid). The equivalent of DNA with deoxyribose replaced by ribose and with the base uracil instead of the rather similar base thymine. *See also* **Transcription**, **messenger RNA**.

Sequence. The precise order of bases in a nucleic acid molecule or of amino acids in a polypeptide.

Somatic cell. A body cell, in contradistinction to a reproductive or germ cell.

Splicing. The joining of cut ends of double-stranded DNA (*see* **ligase**) or of single-stranded RNA, as in the removal of **intron** sequences from **messenger RNA**.

Targeting. The integration of a gene, supplied to the cell as a DNA fragment, into a desired place in the genome, usually the site at which it naturally resides.

Ti (tumour-inducing) plasmid. Present in *Agrobacterium tumefaciens*. A segment (T-DNA) of the plasmid DNA is integrated into the genome of the host plant and is responsible for the transformation of normal cells to tumour cells. With appropriate modifications, the T-DNA can carry foreign DNA sequences of any desired kind into the genome of a susceptible plant. *See also* **Plasmid**.

Transcription. The enzyme-catalysed synthesis of a strand of RNA using a strand of DNA as a template, according to the base-**complementarity** rules.

Transformation. Change in type of a cell or organism brought about by artificial introduction of DNA.

Transgenic. Used to describe an organism carrying a foreign gene as the result of transformation.

Translation. The synthesis of a **polypeptide** chain of a specific amino acid sequence determined by the base sequence in the corresponding **messenger RNA**. *See also* **Ribosome**.

Upstream. Refers to position in DNA relative to the direction of transcription of a gene. The region upstream of the gene **promoter** is often involved in regulation of gene activity according to cell or tissue type and stage of development. *See also* **Downstream**.

Vaccine. A preparation from, or a derivative of, a pathogenic bacterium or virus, used to induce immunity without itself causing disease.

Vector. The DNA construction, often based on a **plasmid** or **virus**, into which a gene is ligated (*see* **Ligase**) for **cloning** (usually in a bacterium) or for integration into a **genome**. Some vectors (expression vectors) are especially designed for promotion of gene transcription.

Virus. An infective agent, based either on DNA or on RNA, that can get itself replicated in host cells, but with no cellular structure of its own, relying on the host cell for most of the metabolic functions needed for its own propagation. *See also* **Bacteriophage**.

Walking. In the genetic context, the analysis of the sequence of DNA segments along long stretches of chromosome by **cloning** of overlapping restriction fragments (*see* **RFLP**), each one being used as a probe for the next.

INDEX